THE NEW SⅭⅠⅠⅠⅼ
CONSCIOUSNESS SURVIVAL
And The Metaparadigm Shift
To A Conscious Universe

By Alan Ross Hugenot

First published by Dog Ear Publishing
4011 Vincennes Rd
Indianapolis, IN 46268
www.dogearpublishing.net

dog ear
PUBLISHING

ISBN: 978-1-4575-4694-5

This book is printed on acid-free paper.

Printed in the United States of America

FOREWORD

Celebrating Consciousness in
Post-Materialist Science, Spirituality, and Society[1,2]

Gary E. Schwartz, Ph.D.[3]

[1]FOREWORD for Dr. Alan Hugenot's ***THE NEW SCIENCE OF***
CONSCIOUSNESS SURVIVAL *AND THE META-PARADIGM SHIFT*
TO A CONSCIOUS UNIVERSE.

[2]I wish to thank Rhonda Eklund Schwartz for her valuable editorial suggestions on earlier drafts of this Foreword.

[2]Gary E. Schwartz, Ph.D., Professor of Psychology, Medicine, Neu-
rology, Psychiatry, and Surgery
Director, Laboratory for Advances in Consciousness and Health,
Department of Psychology, The University of Arizona
1601 N. Tucson Blvd., Medical Square Suite 17, Tucson, AZ 85716:
Email: gschwart@u.arizona.edu

Introduction: Two Warnings and the Mind-Material Relationship

How often do we read a book—let alone two books—beginning with a stern
warning from its respective authors?

Let me quote a warning I happened to write for the Preface to a book titled
AN ATHEIST IN HEAVEN: Evidence for Life after Death from a Lifelong Skeptic. I
share this warning not only because it (1) also accurately describes the book
you are now reading, but because (2) Dr. Hugenot happened to write an
even stronger warning for this book.

Here is the warning I wrote:

WARNING—THE BOOK YOU ARE ABOUT TO READ IS ABSOLUTELY TRUE

This may sound like an odd warning, and I can state for a fact that I have never read
(or written) such a warning before. However, in this instance, the warning is advisable

and completely justified. What you are about to read is factually true, historically accurate, and should be taken very seriously.

As my warning implies, I had not (yet) read a draft of Dr. Hugenot's book including his warning, and the reverse was true as well—he had not read mine. I decided to share my warning here not as an advertisement but as an important advisement for Dr. Hugenot's brilliant book.

I quote Dr. Hugenot's warning below because it deserves special emphasis. He wrote:

WARNING - PHILOSOPHICAL CAUTION IS ADVISED:

*This book may not be suitable for all audiences **because it champions real science instead of pseudoscience**. Further, it also does not champion nor conform to any religious views. Instead, it speaks of scientifically verified consciousness survival in an afterlife. Consequently, stalwart materialists and religious persons of any sect will find its conclusions disagree with their chosen Belief Systems (their BS).*

As you can see, Dr. Hugenot is not pulling any punches in this book. He is telling it as he sees it, as concisely and forthrightly as he can. He is on a mission—to awaken humanity to the emerging revolution in contemporary science which sees consciousness as primary—not secondary—to the physical existence of nature and the universe, and this includes the primacy of human consciousness and its survival beyond physical death.

Just as the Copernican Revolution taught us that we had it backward about the relationship between the sun and the earth, the Post-Materialist Revolution is teaching us that we also have it backward about the relationship between consciousness and the material brain. Just as the sun does not revolve around the earth, the origin of consciousness (and mind more broadly) is not created by the material brain.

Just as our commonsense view that the sun is very small compared to the huge size / prominence of the earth is wrong (in fact, it is the reverse), our commonsense view that our mind is very small compared to huge complexity / prominence of the brain is wrong (it is also the reverse).

Instead, just as the earth is ultimately secondary to the sun, the material brain (and matter in general) appears to be secondary to the mind.

Mind is primary, material is secondary. This statement is as simple as $e = mc^2$; easy to read and remember.

This simple idea, originally formulated by ancient thinkers, mystics, and shamans, is being resurrected and reframed by contemporary science.

Although most of us are not aware of this yet, this Post-Materialist Science Revolution makes the Copernican Revolution seem like child's play.

I have a special interest in consciousness and the emergence of the Post-Materialist Science Revolution. In fact, I hosted and co-organized with Dr. Mario Beauregard from the University of Arizona and Dr. Lisa Miller from Columbia University an International Summit on Post-Materialist Science, Spirituality, and Society. Held at Canyon Ranch in Tucson, Arizona, in February 2014, this summit resulted in writing of a Manifesto for Post-Materialist Science (http://www.explorejournal.com/article/S1550-8307%2814%2900116-5/fulltext) and the creation of a website featuring consciousness and post-materialist science (see www.opensciences.org). *Editor's Note: The manifesto is also quoted in Chapter 2.*

Dr. Hugenot's metaparadigm vision is central to this emerging revolution.

Who is Dr. Hugenot? "Fighting fire with fire" for the Post-Materialist Science Revolution
Before considering a few of Dr. Hugenot's honest and candid statements about mainstream science and its erroneous views about consciousness, we should understand who Dr. Hugenot is.

Dr. Hugenot is a highly successful engineer and applied scientist specializing in naval architecture and marine engineering. He understands sophisticated marine engines and is gifted at problem solving when these huge and complex machines malfunction. Dr. Hugenot is a Navy man. He testifies in contentious court cases, and he appreciates both scientific and legal evidence.

At the same time, Dr. Hugenot is a deeply spiritual person who has become a well-educated and highly skilled evidential (evidence-based) medium. He knows through direct personal experience that at least some mediums are real (i.e., they are not frauds or mental magicians), and he is committed to proving this once and for all.

Over the course of more than fifteen years, I have had the privilege of experimentally testing and working closely with approximately thirty genuine evidential mediums. These mediums have participated in single-blinded, double-blinded, and/or triple-blinded controlled experiments, and passed these critical evidence-based tests.

Most of these gifted, evidence-oriented mediums have authored books about their experiences, and many have written multiple books. Some have become quite famous, appearing on well-known television shows (e.g., *The Today Show* and *Dr. Oz*) and even having their own TV shows (e.g., *Crossing Over with John Edward*). Only a few have earned advanced doctoral degrees (e.g., in psychology, medicine, and law). But to the best of my knowledge, only one has advanced doctoral training in the physical and engineering sciences, and this is Dr. Hugenot.

I have come to know Dr. Hugenot, and it is obvious that he loves the power and beauty of the scientific method. He is also a truth seeker in the tradition of great scientists, and his heroes represent a who's who in the history of science.

At the same time, Dr. Hugenot does not suffer fools lightly, and he is angry with a subset of (1) dogmatic materialist **scientists** who dismiss contemporary consciousness science (especially survival of consciousness and psi research), as well as with (2) dogmatic materialist **skeptics** who dismiss this research as well. It is worth noting that a sizeable number of individuals wear both of these professional and conceptually biased hats (for example, not all skeptics are professionals who are skeptics; in fact, many are not scientists).

Sometimes it is necessary to **"fight fire with fire"** and Dr. Hugenot has decided to become, metaphorically, what I am calling a "firefighter" for the Post-Materialist Revolution. For example, he writes:

> *Today the materialist scientific community fully admits the professional hubris of having completely failed to discern 96 percent of the now known universe.*
>
> *This is obvious from their acceptance of the proof for dark energy and dark matter, which make up 96 percent of the universe, but which our best materialist science cannot yet discern. If their best efforts looking inside the box missed 96 percent of everything…could it be that maybe their box is too small?*
>
> *But these same "scientists" (and I allow them that title only guardedly), whilst admitting that they don't know 96 percent of what there is to know, will still naively insist that "We know consciousness survival is impossible" and "We know that psi is impossible."*
>
> *They are actually saying, "We only know four percent of what there is to know, but we're sure that the ninety-six percent we don't know does not include those things we don't particularly like." Now, if that sounds a bit childish…that is because only an immature intellect can function with such cognitive dissonance. One has to ask, "How can they be that stupid?" Only if materialism is their religion.*

Those are fighting words.

For the record, unlike Dr. Hugenot, I am not a "firefighter." Dr. Hugenot was a Boy Scout who then became a Sea Scout—in fact, he earned the very rare award of becoming a "Quartermaster Sea Scout." This is the equivalent of becoming an advanced Eagle Scout only it is much more exclusive; you can think of Quartermaster Sea Scouts as seagoing Eagle Scouts. In his adult life Dr. Hugenot became "Fleet Leadership Training Officer" for the Northwest Area of the Boy Scouts of America, which included Oregon, Washington and Idahoe.

Dr. Hugenot is an avid sailor and boatman (his latest boat is a 47 ft schooner named SEA RAVEN), and he received a sea captain's license. He is also rugged, strong, has great posture, and stands well over six feet tall.

In contrast, I never made it through Cub Scouts, my early passions did not require ruggedness or physical strength (e.g., I played professional guitar for ten years in high school and college), and when I remember to stand straight, I measure maybe five foot seven. Though I did serve a year in Army ROTC when I was a first-year electrical engineering student at Cornell University, my professional service has been as an academic professor whose career has spanned Harvard University, Yale University, and the University of Arizona. My role over the years has been to provide critical evidence which, metaphorically speaking, supports the "essential role of firefighters who are squelching the dangerous flames of excessive and erroneous materialist rhetoric and action." Dr. Hugenot takes the position that it is necessary, so to speak, to actively put out the dangerous fires of excessive and erroneous BS.

I am writing this Foreword to express my enthusiastic support and gratitude for Dr. Hugenot and his work, not only for this highly readable integrative celebration of emerging consciousness theory and applications to life beyond death, but for his bravery and integrity in forcefully expressing what he sees as being the correct path for human understanding and evolution in these areas.

The Lesson of Eagle Science versus Ostrich Science

Instead of attempting to highlight some of Dr. Hugenot's core insights and conclusions about the physics of consciousness, the nature of the human spirit and soul, and its implications for science and society (best you learn this directly from him), I thought it might be valuable to help you understand Dr. Hugenot's philosophical approach to science and life via a framework I developed which is both metaphorical and practical (as well as spiritual).

I refer to this metaconceptual framework as the distinction between Eagle Science and Ostrich Science. It is based on both physical and metaphorical distinctions associated with these two great birds, and it seems appropriate to share this lesson in a metaparadigm book.

What I have learned over the years is that many of us in general, and scientists in particular, are trained to practice what can be called "Ostrich Thinking." The ostrich is a magnificent creature in both its novelty as well as its strength. We often associate ostriches with putting their "heads in the sand"—a simplistic coping mechanism for avoiding seeing potential threats (which may be psychologically comforting but is potentially dangerous, especially if the threats are serious).

In all fairness to these special birds, ostriches do not actually use this as a coping mechanism; they mostly put their heads in the sand to attend to their infant chicks.

We scientists are typically taught to put our "minds in the sand" in many ways. We are taught to specialize, to restrict our curiosity, to limit our attention to the details of our respective research areas, and to treat novel or seemingly anomalous information as being suspect (often to be ignored at all costs unless absolutely required). This is both a gift and a curse.

The positives are well known; it makes science manageable and practical, it encourages scientists to be careful and thorough, and it makes it feasible for scientists to have careers and support themselves and their families.

However, the negatives are that it discourages creativity and openness to new information, especially if the information is experienced as controversial or anomalous. It encourages scientists not only to miss the forest for the trees, but even to miss the trees for the branches and the twigs.

Though ostriches stand tall (and they run very fast and powerfully), they do not fly. They are restricted to seeing the world entirely from the perspective of standing on the ground.

In contrast, consider the eagle and what I call "Eagle Thinking." Eagles have two qualities which are in sharp contrast to ostriches. First is that eagles fly very high—they are among the highest flying birds. From this great vantage point, they can see "the big picture"—they can see not only the forests but the tops of mountain ranges and the depths of the valleys as well. In contrast to ostrich vision, eagle vision is big-picture vision.

At the same time, eagles see the details. They can see tiny rabbits miles away and thereby feed themselves and their offspring. Eagles not only see the forests (the whole), they see the trees, branches, and twigs (parts) as well. This is sometimes called the "eagle eye."

I have a special fondness for birds, and I love tiny hummingbirds and finches as well as large ostriches and eagles. Each bird is unique; each has its own special beauty and abilities as well as its respective limitations. In the grand scheme of things, ostrich thinking and eagle thinking both have their merits and value. In fact, if styles of thinking were given "merit badges" for Boy Scouts or Girl Scouts, one might wish to earn merit badges in both.

However, *when the specific challenges facing us require innovative big-picture thinking which sees the whole as well as the parts, there is a great advantage in having eagle vision.*

As mentioned previously, Dr. Hugenot was a seagoing Eagle Scout as an adolescent, and this propensity and training probably nurtured his high-level eagle vision today. The book you are reading was not written to be a comprehensive, scholarly, historical treatise of the mind-matter relationship and the survival of consciousness beyond physical death (a number of excellent academic books of this sort currently exist). Instead, Dr. Hugenot has intentionally written a personal, integrative, no-holds-barred, essential, and inspiring vision for the average reader about the new science of consciousness survival.

This book is already historic. It is the first book written by someone who is both a scientist and a medium. If anyone is going to successfully integrate and advance future science and our understanding of consciousness and its survival, it is Dr. Hugenot. Science, spirituality, and society have long awaited this bold and beautiful book.

PREFACE:

"If it could be shown that there is truth to any of these notions (such as telekinesis, astrology, precognition, mediumship, channeling and clairvoyance as 'would-be' sciences), it would be the discovery of the century, much more exciting and important than anything going on in the normal work of physics." [1]
                    Steven Weinberg, Nobel Laureate in Physics

"Ether or Aether, a material substance of a more subtle kind than visible bodies, supposed to exist in those parts of space which are apparently empty." [2]
                    James Clerk Maxwell, 9th ed. Encyclopedia Britannica

"It is the responsibility of scientists never to suppress knowledge, no matter how awkward that knowledge is, no matter how it may bother those in power. We are not smart enough to decide which pieces of knowledge are permissible and which are not." [3]
                    Dr. Carl Sagan (1934–1996)

"For me, it is far better to grasp the Universe as it really is than to persist in delusion, however satisfying and reassuring."
                    Dr. Carl Sagan (1934–1996)

"Alan Hugenot lectures on 'the Leading-Edge Science of the Afterlife'... he concluded that the entire universe is conscious and that this explains both near-death experiences and certain paradoxes of quantum theory... As someone with a physics degree, I know that Hugenot's...basic idea of a conscious universe is neither crazy nor new... Erwin Schrödinger, one of the fathers of quantum physics, was an avid student of Hindu philosophy, and believed something similar."
                    Gideon Lichfield, April 2015 Atlantic Monthly

TO BEGIN WITH:

For nearly a century the die-hard *Materialists* have attacked parapsychology, trying to force the battle to be waged in proving whether or not psi (remote viewing, ESP, psycho-kinesis, etc.) is real or not, and they do this while knowing that the foundations of their *Materialist* paradigm have been entirely disproven by modern science (quantum electro-dynamics and non-locality). And while also seemingly ignoring the fact that their own "god-less" theories, require a couple of "creation miracles" far more amazing than those claimed by most religions.

Instead, it is time for these amazingly shallow thinking religious zealots of *Materialism* to either grow up or be left on the wrong side of history, remembered as the religious fools they actually are. This book is intended to set them on the defensive by moving the battle over to them, and making them defend their myopic, recalcitrant adherence to a fully disproven belief system (BS).

It is time for *Materialists* to do as Dr. Carl Sagan advised and provide the **"extraordinary proof, for their extraordinary claims"** and explain (to the more rational scientific community) why they hold such insane adherence to a dead philosophy that is completely disproven by modern science. And further, why are they so afraid of any proof of consciousness survival?

Instead of their loud declamations against all parapsychological data, and their trademark character assassination of all paranormal scientists, they instead need to show their own data proving *Materialism*... But there never has been any such data, which is why they so vehemently ostracize anyone who disputes their falsehoods... And further, where is their use of the scientific method? The method, which they so vociferously claim to be following but, and ignore entirely, relying on innuendo and slander to make their points. Frankly, *the emperor simply has no clothes.*

There are many pseudoscientists who (exactly like the church fathers of the seventeenth century) now see themselves as *the newly self-proclaimed purveyors of truth*; they have replaced the church with a new religion of *Materialism*. So it seems that one truth continues, as stated by Jonathan Swift:

> *"When a true genius appears in this world, you may know him by this sign, that the dunces are all in confederacy against him."*

Today, the *Materialist* scientific community grudgingly admits the professional hubris[4] of having completely failed to discern 96 percent of the now known universe. This "oversight" is obvious, from their acceptance of the proof for *Dark Energy* and *Dark Matter*, which by all calculations make up 96 percent of the universe... Yet application of the best *Materialist* science cannot yet discern what this 96 percent even is.

Frankly, if their best efforts looking inside the Newtonian box missed 96 percent of everything...could it possibly be that maybe their box is too small and there is a good deal which was left outside the box?

Recently, I heard one vocal *Materialist* zealot (a professor at a prominent California technical institute) vehemently say in a taped lecture, "*I can find no place within the standard model for consciousness to exist.*" Indeed, is this so-called professor completely unaware that Rene Descartes and Isaac Newton purposefully placed consciousness outside that box 350 years ago?

Apparently he is actually that naïve. Although he holds a Harvard PhD, this remark indicates that he is either totally unaware of that "trivial" (but important) bit of scientific history, or he is hiding from the truth. Really, you can't seriously expect to find something in the box when the founding fathers purposefully left it outside the box, unless you are just completely ignorant of an historical fact most others do know.

Yet modern *Materialist* "scientists" (and I allow them that title only guardedly), while admitting that they don't know 96 percent of what there is to know in the universe, will still naively insist *"We do know that consciousness survival is impossible"* and *"We do know that psi is impossible."*

This irrational, cognitive dissonance results in a belief that *"We only know four percent of what there is to know, but we can be sure that the ninety-six percent we don't know cannot possibly include those things we don't particularly like."* Now, if that sounds a bit immature and childish, that is because only an immature intellect can function with such massive cognitive dissonance.

One then has to ask, **"How can they be that stupid?"**

The short answer is *"Only if Materialism is their religion,"* which in their zeal for their BS (belief system) to be true skews their perspective.

Such absurdity (based entirely on belief and no evidence) amounts to no more than a religious dogma, which is why throughout this book I carefully capitalize the word *Materialism* as an editorial courtesy that I provide for all religions (Christians, Jews, Muslims, Materialists, and Atheists). Yes, those last two are also religions.

Frankly, such a catastrophic failure to discern 96 percent of the universe is precisely because of their arrogant rejection of the paranormal (psi) and their continuing zealous religious refusal to look outside their self-imposed box, to examine the plentiful data for psi which has been repeatedly replicated in over 130 years of careful parapsychological science (formerly psychic science).

Science has never been a set of rules carved in stone, nor is it a set of doctrines and dogmas written in a Bible or a Quran. Science is a mental process, which requires a mature intellectual honesty capable of looking at <u>all</u> the data (even conflicting data which may disprove your own pet theory) and then true science requires adjusting your pet theories and underlying hypothesis to suit that new, but conflicting data.

Simply because data is "uncomfortable" and you *choose* to see the paranormal as "impossible" is not science…but it is definitely childish, and it is the exercise of mere BS. Yet *Materialists* foolishly refuse to examine the evidence for psi, no matter how carefully controlled the experiments and no matter how well credentialed the parapsychological scientists who did the experiments. Such automatic rejection of all things paranormal is irrational. But *Materialists* blindly and religiously continue to do this and so quite simply do not to discern 96 percent of the universe. Duh?

Today (in 2016), the existing scientific establishment (made up mostly of *Materialists*) has more egg on their face than did the Vatican back in 1633 for refusing to look through Galileo's telescope. It is only the parapsychological frontier scientists who don't have egg on their faces.

In this book, I have shown *Materialism* to be the archaic superstition that it actually is. I have given the URL links to scientific and medical videos that back up all of the findings of parapsychological science and also references to all the scientific literature that shows the paranormal is not only real, but that it is truly the "normal" and that *Materialism* itself is truly the paranormal, or more correctly the abnormal.

And I specifically call out these pseudoscience experts challenging all the *Materialist* skeptics (Shermer, Alcock, Blackmore, Randi, et al) to try real science, instead of arguing with me and assassinating my character with innuendos and lies, calling me and the other scientists frauds, as these *non-scientist psychologists* have always done to any reputable scientist.

But as Gideon Lichfield pointed out with his statement about my theories in the *Atlantic Monthly* (April 2015), when *Materialist* skeptics attack me, they are really attacking **Erwin Schrödinger, Max Planck, Werner Heisenberg, David Bohm, Albert Einstein, Paul Dirac, Wolfgang Pauli, Ernest Rutherford**, and all the other founders of quantum electrodynamics. Instead, it is time to quit being immature, irrational children and grow up.

If they would expend the same anger energy doing their homework, stop kissing up to the "scientific method" and instead actually use it, while no longer vilifying honest frontier scientists who are using that scientific method, and stop merely assuming the paranormal is "impossible," then they would shortly realize *the party is over and Materialism is dead.* (But maybe that is exactly what they are so afraid of.) It is time for any intelligent *Materialist,* to become a real scientist and actually start looking at all the repeatedly replicated data proving the paranormal is normal. Then modify your paradigm to suit the truth, and get on the right side of history or be left in the fool's bin of stupidity along with the religious hierarchy of the sixteenth and seventeenth centuries.

Suddenly, instead of laughing at the parapsychologists (which *Materialist* zealots have done since the McCarthy era), everybody will now be laughing at them for stupidly clinging to the religion of *Materialism* as if it were any more "true" than other superstitions.

Many of them will just cross the street and say, *"Oh yeah, I always knew that"* about the conscious universe.

IT'S SIMPLY TIME TO GROW UP AND ACT LIKE ADULTS: While the older calcified skeptics and the younger more foolish *Materialists* will continue to fight for their religion, intelligent scientists need to heed Dr. Sagan's advice, by no longer stupidly rejecting out of hand the science of psi merely because they choose the foolish religious superstition of "believing" it is impossible.

Former arrogant *Materialists* also need to apologize to the honest parapsychological frontier scientists who have, despite rejection by the ivory towers, persisted in actually investigating psi. Including an apology for the ridicule they derisively showered upon these more intelligent scientists (smart enough to study psi) whom they previously ostracized as they religiously tried to "kill the infidels."

After choosing to act maturely, reopening their minds to real science, and taking off their chosen blinders (of *Materialism*), they will quickly discover that this so-called paranormal is more normal than their cherished illusion of a physical reality. Shortly, they will discover that an invisible, *implicate* field of consciousness exists entirely outside the limited perceptive boundaries defining the material world (the Newtonian box), and further that this *implicate* field provides the matrix for the visible, *explicate* reality of our physical world of matter. And they can join the scientific "gold rush" to discover how it all works. If they maintain their religious views, they will miss this great opportunity.

This *implicate* field is located in an un-manifested state of quantum potential, which exists simultaneously with our observed reality, either in alternative dimensions of *Dark Matter* and *Dark Energy* or within non-visible higher or lower frequencies of the electromagnetic spectrum retained in plasma fields located within *non-local* space. Current scientific investigations by the physicists working in "the void" ("the vacuum" or zero-point field, ZPF) are figuring this out. Additionally, many cell biologists have already discovered the same consciousness matrix to exist. This new *biocentric* (life-centered) perspective from biology is fully compatible with the modern physics of quantum electrodynamics (QED). And further, QED and this new *biocentric* perspective do not conflict with each other, and both fully allow scientifically for consciousness survival in an afterlife.

MATERIALISM IS SIMPLY DEAD SCIENCE (excuse the pun): Not surprisingly, because science is a process and not a set of "commandments," the prevailing (yet archaic, 350-year-old) *Materialist* perspective, which much of Western culture takes for truth, has been completely falsified by the proven validity of QED and *non-locality*, and intelligent modern scientists have known this for nearly ninety years with QED, and since 1972 (forty-four years) with non-locality.

Yet despite the fact that *Materialism* is no longer at all compatible with modern physics, it remains a dogmatic article of faith, forcefully taught by the current crop of lockstep post-docs, who (like medieval monks) unwittingly carry on the biased religious beliefs of their university mentors (high priests) even though *Materialism* is merely an archaic, calcified superstition completely disproven by QED.

Unfortunately, while the need for a conscious universe to form the underlying matrix of our physical reality was known by the Berlin physics intelligentsia even as early as 1900, this prior knowledge has been lost to us in Western science for more than a century.

THE BLINDNESS OF WESTERN SCIENCE: This militant blindness in American science was caused by an unfortunate twist of history. The "German Geniuses" (Planck, Einstein, Heisenberg, de Broglie, Schrödinger, Bohr, Pauli, Dirac, not all of whom were actually German, but their center was in Berlin) were leading the scientific learning of the world, and had (prior to 1914) discovered that the universe was conscious. But this intellectual advance was interrupted by two world wars, and discredited in America and the West due to the repugnance the world felt for all things "German" after the discovery of the holocaust.

Consequently, international physics lost the intellectual leadership of these brilliant, but German, scientists, that is except for Einstein, who was the only one of the Berlin group who retained any credibility, and only because, as a Jew, he fled Berlin in 1938 and joined the faculty at Princeton.

But unfortunately, after the Americanized Germans (Werner Von Braun and Albert Einstein) passed on, and the gifted scientists David Bohm and J. Robert Oppenheimer were discredited and pushed aside by the McCarthy witch hunters, the remaining American physicists who moved into the intellectual vacuum left in international physics were themselves only "second level" intellects. These lesser intellects understood *Materialism*, but could not make the quantum leap in thinking required to comprehend the consciousness implications proven by QED. Except for one or two rare exceptions (Feynman, et al) these American "leaders" were not Nobel laureates, not independent thinkers; they were able to follow but not lead. Yet because of the ancient rule which applies in all institutions of higher learning, ***"Those who can do, and those who can't...teach,"*** these lesser thinkers became the mentors for our current crop of post-doc *Materialists* who still continue in lockstep.

AMERICAN SCIENCE HAS NEVER LED THE WORLD: In America, we simply have a lot of bloated pride. We conveniently forget that most of what we produce in Silicon Valley is due to scientific concepts developed in England, Germany, Italy, and the Baltic states, but not in America. Unfortunately, ever since our Yankee ingenuity won the Second World War and we took up the mantle of global leadership given up by the British Empire, most Americans naively think that the USA is the leader in all things.

Yet amusingly, when we needed real scientists to get the job done during the space race (between the US and Russia), we lacked any American scientists with the skills to compete. Consequently, the space race became a contest between brilliant expatriate German physicists in America (Werner Von Braun, etc.) working to beat the brilliant German expatriate physicists in Russia. This contest between the two superpowers to see who would be the first to the moon was actually a contest between two groups of Germans. A beautiful line from James Michener's book *Space* says it clearly: ***"Our Germans were just smarter and faster than their Germans."***

Unfortunately, today, the majority of American high school and college physics and biology professors (*materialist* post-docs) are unaware that they are being trained to miss the boat (literally 96 percent of the universe) by perpetrating the thinking of such "second-level intellect" physicists who rigidly indoctrinated them into their archaic mind-set of belief in the 350-year-old dogmas of *Materialism* (classical Newtonian physics). They naively continue to teach that ***"anything we can't see (like consciousness) is not real."*** I know this personally because I was trained in the Science Engineering & Technology (SET) programs by those same mentors in American colleges and universities.

Here are some helpful definitions:

- **TRUE SCIENCE** is a mental process that requires the mature intellectual honesty to look at all the data (even conflicting data that may disprove your pet theory) and then to adjust your underlying hypothesis to suit that new but conflicting data.

- **PSEUDOSCIENCE** is a dishonest intellectual mental process (widely practiced by *materialists* as well as religious leaders) of choosing to ignore any data that conflicts with your own pet theories. The data is ignored to protect existing assumptions, dogmas, and belief systems (BS).

- **MATERIALISM** is a paradigm (viewpoint) that believes in a physical reality made of matter (substance) but is based solely on a collection of archaic and mistaken presuppositions known as the ***First Principles*** (***reality, locality, causality, continuity, determinism*** and ***certainty***). These First Principles were culturally adopted in the pre-scientific 1600s, prior to the advent of science and later, without investigation, were presume to merely be self-evident. No one later worked out the proof for these presumptions.

- **FIRST PRINCIPLES OF MATERIALISM ALL FALSIFIED BY QED:** Yet all of these presumptive **First Principles** were completely disproven between 1905 and 1935, with the advent of relativity and quantum electrodynamics. Unfortunately, scientists trained in *Materialism* are unable to fathom the implications of our quantum world, and instead hold as dogma their archaic belief system (BS) of *materialism.* Consequently, they commit the hubris of *believing* they already know what is possible and what is impossible; and therefore *a priori* **they unscientifically refuse to look at any conflicting data** which is "known" to be "impossible" within the limited framework of their preconceived *materialist* paradigm.

On the other hand, while this book definitely condemns the religion of *Materialism,* it does not champion or condemn any other recognized religions. In fact, fully comprehending the message of the book will require both those with a *Materialist* belief system (BS) and those with any other religious belief system (BS) to give up several treasured superstitions of their chosen BS. Instead, this book illustrates the proven scientific reality of consciousness survival in a non-religious afterlife. Actually, an atheist's dream, if the reader can see beyond their automatic bias.

1. THE NEW SCIENCE OF CONSCIOUSNESS SURVIVAL

The existence of a hidden field (Dr. David Bohm's *implicate* order) of non-physical consciousness, occupying as yet undiscerned additional dimensions (within *Dark Energy* or etheric plasmas) located outside the visible reality (Bohm's *explicate* order), which is itself limited by our perceived Newtonian box of 3-D plus time, has now been proven to exist scientifically by the following collated data:

- The philosophical implications of quantum electrodynamic theory (QED), requiring a living consciousness separate from the limited paradigm encompassed by 3-D plus time. (See Chapter 11);
- Recent repeated replication of John Bell's theory of non-locality;
- Astronomical observations indicating the existence of undiscerned *Dark Energy* and *Dark matter* making up 96 percent of the universe;
- Studies of the near-death experience (iands.org). (I had a personal NDE in 1970, see Chapter 6);
- Children who remember past lives, University of Virginia (med.virginia.edu/psychiatry), (see Chapter 7);
- After-death communications demonstrated in triple-blind laboratory experiments testing evidential mediumship, which I personally participated in (see Chapter 8);
  1. Consciousness Research Laboratory, Institute of Noetic Sciences (noetic.org)
  2. Laboratory for Advances in Consciousness and Health (LACH) at the University of Arizona (Lach.web.arizona.edu)
  3. Psychical Research Foundation (psychicalresearchfoundation.com)
     - Hypothermic cardiac arrest: (See Chapter 5)
     - Remote viewing (see Chapter 10); and
     - The emerging science of biocentrism (see Chapter 12).

This hidden field (*implicate* order) of non-physical consciousness also provides the matrix upon which the *explicate* order of observed reality is continually manifested, exactly as Dr. Max Planck stated in 1900.

Further, this consciousness as the matrix perspective is fully supported by quantum electrodynamics (QED) non-locality and the biocentric paradigm which, taken together, both see life and consciousness as fundamental to the formation of the universe.

2. DEFINING THE PROBLEM
THE LOST CENTURY: Except for William Shockley's transistor (1947) and the work of Alan Turing and others (1945), which resulted in digital computing, the advancement of international science has been almost completely stalemated since the end of the Victorian era (1901). Yes, we have

miniaturized the solid state circuit down to nano-scales, but even DARPA's Internet is not a new idea. Most people are unaware that a complete analog network existed as early as 1940. The teletype system operated over existing phone lines. The problem with Ma Bell's 1940s mechanical Internet was high cost. A teletype installation cost multiple thousands, back when average salaries were under $2,000 per year. But other than affordability and mass application, the digital Internet is only an incremental improvement on that basic idea and not a "new" idea.

The promise of continual progress that was so clearly felt at the dawn of the twentieth century in Europe, America, and the British Empire—following on the advent of telephones, radio, flight, automobiles, public education, and the international cooperation of British, German, American, French, Danish, Swiss, and Italian scientists and inventors all sharing what they knew to allow the advancement of the whole world and not just their specific national interests (as it existed before the first world war)—was unfortunately lost because this intellectual stirring was centered in Berlin.

This concentration of Nobel laureates in the universities at Berlin had begun to realize that the underlying matrix upon which our reality continually manifests from the void is in fact *consciousness*. But this concentration of super genius was disrupted thanks to the childish egos of two German leaders (twenty years apart), Kaiser Wilhelm and later Adolf Hitler, along with two entirely senseless world wars which these rampant children caused.

Later, after 1945, no one could trust the intellectual integrity of a German scientist, simply because we all imagined the following scenario:

> German citizen: *"Vee didn't know avout da camps."*

> American soldier: *"Okay, so you didn't know... But where the hell did you think Hitler took all the Jews then as he packed them off in those cattle cars? Did you imagine they were taking summer vacations in Bavaria? And did you also not happen to notice the slave labor road crews all wearing striped pajamas with yellow Stars of David attached?"*

The American has now silenced the German citizen.

Yet the smug American soldier forgot to ask himself, *"What would I have done if my country was taken over by a dictator with a gestapo, and if I knew I would be put in the cattle cars also if I said anything at all—would I have spoken out, and gotten myself killed?"*

Righteousness is so easy when you don't face a threat yourself...

Nevertheless, the disruption of the wars stalled the progress of physics and the spreading of the philosophical meaning of quantum electrodynamics (QED). All this learning about the matrix of consciousness which underlies our universe that was comprehended by these Nobel laureate geniuses between 1900 and 1934 was lost to the next generation of American scientists and engineers. What these great European minds had learned was ignored... Their papers were not read; their conclusions were set aside.

Young physicists instead learned the mathematics of the new physics as they raced to make their government the first to establish nuclear war, and later as maturing scientists they continued to work for the military-industrial complex. But they neglected to consider what the implications of this knowledge really meant because they had never been taught to think philosophically.

The advance of true science then slowed down and the rest of what we have seen as "progress" in the latter half of the twentieth century has merely been incremental improvement on the existing technologies.

Even our wonderful cell phones are merely a refinement of the same radio technology which Nikola Tesla (Serbian) and Sir Oliver Lodge (British) demonstrated in the later 1890s and which Marconi (Italian) showed could be used for long-distance transmission in 1902. The digital computer itself is just an outgrowth of mathematical programs and systems invented by IBM and others prior to 1900, and worked out by Alan Turing (British), pressed into service by the needs of war (Enigma project, etc.) which have merely been miniaturized onto silicone chips. But all this seeming advancement of the digital age actually uses transistors, which are entirely dependent on QED being true...and if QED is true it requires that the universe is conscious.

Unfortunately, with the loss of this learned leadership, rather than continuing to move on, physics has foundered in the stagnant pond of *Materialism*, led by lesser minds relying on their archaic (350-year-old) First Principles, which were all completely falsified by QED.

By 1934, quantum electrodynamics (QED) had forced the discovery of the need for a living consciousness to exist entirely outside of physics in order for QED to work. So, the entire Conscious Universe theory was then in place. So what I am saying is nothing new; it has just been ignored and by the majority of *Materialists* conveniently forgotten. This brilliant comprehension drove the founding fathers of QED (Heisenberg, de Broglie, Schrödinger, Pauli, Jeans, and Planck, and their students like Oppenheimer and his student Bohm) into eastern mysticism looking for answers to what the slower and shallower thinking *Materialists* euphemistically termed **"the Measurement Problem"** and its philosophical implications.

Later (in 1965) after that generation of geniuses had all passed on, but with Sin-Itiro Tomonaga, Julian Schwinger and Richard P. Feynman's final mathematical formalization of QED, this problem comes to light (excuse the pun). Now, whenever an intellectually honest scientist examines the data, there is no escaping this requirement for consciousness outside of physics.

THE MEASUREMENT PROBLEM: QED states that in order for Schrödinger's wave function (as expressed in QED) to collapse from potentiality into matter (the illusion of reality that we see), which is illustrated in the double slit experiment (as explained in Chapter 11), a choice is required prior to the collapse into matter (John von Neumann calls this Process 1). This choice must be made by a living consciousness.

Yet this same QED, as Feynman likes to say, is today, *"the most proven theory in all of science"* because it gave us the transistor, digital electronics, cell phones, plasma TV, fMRI scans, etc. If QED is not true, then we wouldn't have these modern electronic conveniences.

Further, recent experimental work at the Institute of Noetic Sciences - IONS (noetic.org) has shown that
"The behavior of quantum objects is exquisitely reactive to the act of observation. The controversial interpretation of this effect, that consciousness is responsible for collapsing the quantum wave function. (John von Neumann's Process 1) has also been supported by other physicists, and there has been much debate. But IONS has taken a more pragmatic approach by explicitly testing von Neumann's ideas. In a recent paper they showed results from three separate experiments all of which showed results consistent with von Neumann's proposal." Series of Experiments Shed Light on the Role of Consciousness, By Dr. Dean Radin, Ph.D. noetic.org/blog/dean-radin/double slip experiment.

IGNORING THE PROBLEM: Unfortunately, as those founding fathers of QED all died off in the late forties and fifties, the electronics revolution (depending entirely on QED being true) was accomplished by engineers and scientists who were trained following the science, engineering and technology (SET) educational programs invented to quickly train graduates during the Second World War and the Cold War which followed, but which (to speed up the curriculum) therefore intentionally excluded philosophy from the curriculum. Today, SET students really need five years of college just to cover the basic SET subjects required for a bachelor's degree in their field; if they were to include philosophy and psychology, it would take six or seven years to get to a BS in any SET field.

Therefore, in the rush to create new graduates needed by industry, SET students are not currently required to study philosophy and the "soft sciences." They are never given the tools to grasp the philosophical implications of

QED, tools that the founding fathers (German geniuses) with more liberal educations did have. Consequently, America has two generations of trained but uneducated scientists, engineers, and technicians.

Today's SET graduates (and I know because I am one) are employed by companies driven by the profit motive, which totally ignores the disturbing fact that there must be a consciousness existing outside of physics choosing the reality that will result (Process 1). This anomalous "fact" does not agree with the widely held *Materialist* paradigm that "reality" is made of substance, so it is easy to simply ignore and hurry on to profit from the digital revolution.

Yet now after the fall of the Berlin Wall and the end of the Cold War, the inventions of high technology have so thoroughly proven QED that modern science has reached an untenable impasse. The philosophical implications can no longer be ignored because the *Materialist* worldview itself has become so distorted (by trying to avoid the increasing anomalies) that nothing further can be understood or learned in science until consciousness is brought into the scientific tent.

Those scientists who are unaware of the philosophical implications of QED, and who still wish to retain the nineteenth century *Materialist* viewpoint that the universe is made of substance, desperately want consciousness to be proven to be an emerging property of the brain. This follows from their unprovable and nonsensical theory that ***"inert matter somehow evolved into conscious matter."***

> ***"One can no more hope to find consciousness by digging in the brain than one can find gravity by digging in the earth."*** Dr. Karl Pribram

Now, let's get real here…how does that work exactly? An unconscious, inert object, say, a stone (we're all made of dust) spontaneously evolves into a consciousness? We didn't fall off the turnip truck yesterday and honestly, even Jesus couldn't do that! So, it appears then that materialism is based on a creation miracle…like ***"Stones gain consciousness,"*** a miracle performed by "Nature" instead of "God." But it sure sounds like religion based on faith in this creation miracle.

Want to stump a *Materialist?* Just ask them, ***"What was here two weeks before the big bang?"*** and you run into the same situation of them having no viable answer… They will eventually answer, ***"Well, it just happened."*** Apparently, the big bang was also a creation miracle.

It is pretty clear, when one takes the time to look into it, that the entire *Materialist* house of cards is built solely on BS (belief systems).

On the other hand, if consciousness is in fact the matrix of all reality, as Max Planck stated in 1900, and as QED requires, then that supposedly dead "matter" (which miraculously became conscious in the *Materialist* paradigm) requires no God-given miracle. Instead, it was simply "always conscious" since the beginning of the universe, and so there is no longer a need for the distortion of an unprovable theory to explain it. Everything in the universe is simply an aspect of consciousness. Ockham's razor says that the simplest explanation is the most logical one… Hello?

It is time for a metaparadigm shift in our cultural scientific worldview, similar to the Copernican revolution or the one which took place in biology with Darwin's theory of evolution. But this shift involves the recognition that consciousness (life) is the matrix upon which the universe is constructed. The recognition that matter is not made of matter (substance), but that at its basic level matter is constructed of consciousness. For this shift to occur each individual must make up his/her own mind by fully understanding the following topics:

- **Science** is the **process** of editing our worldview to fit all the empirical and experimental data.
- **Materialism** is the seemingly self-evident concept that everything is made of substance (matter), and that there is an objective reality "out there" as experienced by our five senses. This perspective was developed 350 years ago using classical Newtonian Physics, based on several foundational concepts which appeared to be "self-evident." This paradigm has become the underlying basis of our western cultural worldview (or metaparadigm). Yet what merely *seems* self-evident is not science. Instead it is merely unproven supposition. Consequently, materialism itself is merely a belief system (BS), literally a chosen religion. Further, since 1934, materialism, including all its foundational presuppositions, has been entirely disproven by quantum electrodynamics (QED).
- **Religion**, on the other hand, is the process of choosing a belief system (BS) and then purposefully ignoring any scientific data that does not fit into our chosen worldview (BS). All superstitions are beliefs without evidence. By this definition, materialism is easily seen to be a religion rather than science.
- **Pseudoscience** pretends to be science but is a mixture of science and religion. This muddled thinking is often expressed by self-styled "scientific" experts who ignore significant scientific data, through the hubris of thinking to know more than he/she actually does, or because he/she believes the thing the data represents is "impossible." But this is really practicing a religion and is not science at all.

- **Quantum electrodynamics** is the latest science of quantum field theory, which best explains our "reality" but which also completely falsifies the underlying foundational presuppositions of materialism.
- **Biocentrism** is an emerging scientific concept that replaces materialism and is also completely compatible with the science of quantum electrodynamics.
- **Consciousness survival** in a conscious universe is an outgrowth of biocentrism that is also fully compatible with quantum electrodynamics. This is an evolving perspective that provides the needed metaparadigm shift.

3. ACTUALLY, I STARTED OUT LIKE ALL THE OTHERS

I was an SET student attending college after serving in the Navy during the Vietnam War. Like the others, I was learning how to work physics problems without thinking about what this all means. But while I was still an undergraduate, suddenly everything changed.

The best thing that ever happened to me was when I died in a motorcycle wreck...

Although that statement may be a bit shocking, it is the most true thing that I know. It forever changed my perspective. In May 1970, I had what is today called a near-death experience (NDE), but at the time that term was unknown. Indeed, it was to me and still remains an actual death, with my return to the body being a reincarnation for a second lifetime in the same physical body.

The entire story is laid out in Chapter 6, but briefly, I was severely injured in a motorcycle accident, lapsed into a coma for twelve hours, and traveled out-of-body, where I communed on the other side with a *Being of Light*. After returning to the body and regaining consciousness, I remained hospitalized for thirty-three days.

But my NDE also took place five years before anyone had heard of what we now call a near-death experience. Dr. Raymond Moody M.D. published his book *Life After Life* in 1975, in which he coined the term *near-death experience*. But back in those dark ages the standard medical procedure was to treat all NDEs as delusions. Consequently, the resident psychiatrist, after attempting to reason with me but finding he was unable to convince me of my error, decided that if I would not conform to his *Materialist* version of reality, he would commit me to the insane asylum, to correct my insanity. This "punishment" was for merely making such insane claims about *having died and then come back to life.* Claims that today are understood by much of the medical establishment as being valid.

One of the questions that psychiatrist dinosaur asked me was:
"Just where do you think such a place as the afterlife could exist?"

I didn't help my case any by responding, *"Doctor, I respect all your learning and degrees, but honestly it is like I've been to Mexico and you haven't, and when I want to tell you about my experience, instead of listening to me and discovering a new country, you are telling me that based on your superior knowledge and training, and although you have never journeyed there, you are sure that Mexico is 'impossible' and I must therefore be mistaken…isn't that a little closed-minded?"*

Luckily, my orthopedic surgeon was not so backward in his medical philosophy, and so discharged me from the hospital a few days early, just ahead of the psychiatrist completing the papers to have me committed.

SO, I BECAME AN ANOMALY: When an event or experience occurs that has no explanation within the current scientific framework, it is considered an *anomaly*. So, by being killed and coming back I had myself become a scientific anomaly. Yet it is such anomalies, which stubbornly refuse to go away, that eventually act as the catalyst for a new advance in science.

Today, forty-six years later, dedicated frontier researchers have proven the NDE as a scientific fact. So, instead of being considered insane, I am now welcomed as a featured speaker. The lesson I learned while out-of-body was that *we are not physical beings, but are instead eternal spirits temporarily occupying physical bodies.*

In a larger sense, I had also learned that our 350-year-old *Materialist* paradigm of classical Newtonian physics, limited to three dimensions plus time, did not include everything. In fact, it fell far short and only included a very small corner of a much larger universe. I realized *Materialist* science was deeply flawed in its worldview.

Burdened with visceral personal knowledge of an inconvenient truth, and being unable to reconcile it with society's standard model of reality, I quickly learned to be quiet about what I knew. Instead, and unlike the other SET students, besides my day classes in science and technology, I immediately began to also study philosophy, history, logic, and metaphysics in evening classes, beginning a research odyssey that has spanned forty-six years investigating multiple scientific disciplines, collating data, and verifying the science supporting the vision presented here, eventually joining the International Association for Near Death Studies (IANDS.org), the Institute of Noetic Sciences (noetic.org), The Academy of Spiritual & Consciousness Studies (ASCSi.org), and the Institute for the Advancement of Science and Consciousness (I-ASC.org, formerly FMBR).

DARK ENERGY & DARK MATTER: Then in 1998, twenty-eight years after my NDE, NASA deployed the Hubble space telescope. For the first time astronomers could look into distant space. As they studied very distant supernovas they repeatedly noticed another *anomaly*. It was apparent, by all calculations, that a long time ago, the universe was not expanding as rapidly as it is today. This meant that *the rate of expansion of the universe was speeding up rather than slowing down due to gravity*, as everyone had previously thought. Now they needed to figure out what was causing the universe to expand, and to accelerate in that expansion, and also revise their understanding of gravity.

Today, twenty years later, *the most plausible way to explain this expansion anomaly is the Dark Energy - Dark Matter hypothesis,* which accounts for this accelerating expansion and is now accepted by most astronomers, cosmologists, and physicists as a fact. *Dark energy* and *dark matter* appear to be proven by three independent observations:
- motion of galaxies,
- structure simulations, and
- temperature fluctuations in the cosmic microwave background.

Consequently, physicists and astronomers are trying to understand just what the dark energy and dark matter are made of and how they are distributed throughout the universe. Attempting to "measure" dark energy, they have observed an additional anomaly which shows that dark matter is actually there, known as *gravitational lensing,* which is the measurement of how much the passage of starlight is "bent" when it passes around a concentrated mass on its way to earth, where we observe it. The greater the bend, the more mass must be present. From these anomalous observations they have calculated that the dark energy (72 percent) and the dark matter (24 percent) make up 96 percent of the universe.

But within their acceptance of the Dark Energy - Dark Matter hypothesis, there is also a dark rider… If, as they theorize, dark energy and dark matter are real, then:
> *Our best science has failed to find 96 percent of the universe!*

Not only is this a colossal failure on the part of *Materialist* science, which illustrates the limitations of that current scientific perspective, but this missing 96 percent gives an answer to that psychiatrist's question: *"Just where do you think such a place as the afterlife could exist?"*

When we can't discern 96% of what exists in our universe, there is plenty of spare room for all kinds of unknowns. Much more than just the afterlife's **"undiscovered country from whose bourn no traveler returns,"** but entire undiscovered galaxies, alternative dimensions, multi-verses, etc. etc.

These accumulating anomalies are forcing a major revolution in our cultural worldview. Scientifically trained researchers across many disciplines (physics, biology, medicine, neuroscience, consciousness) are leading this search (gold rush) for the new metaparadigm, which includes research into consciousness, spirituality, psi, and all things paranormal.

Hopefully, this book will make all this easily understandable in a simple way, showing how *Materialism* has always been merely speculation based on several untrue presuppositions, has now been completely debunked, and also how the evolving scientific worldview, based on quantum electrodynamics, allows the existence of psi, the paranormal, and consciousness survival, to all be valid.

"We are intelligent beings, and intelligent beings could not have been formed by a blind brute insensible thing."[16] Voltaire (Francoise-Marie Arouet 1694–1778)

TABLE OF CONTENTS

Chapter 1

INTRODUCTION

"All matter originates and exists only by virtue of a force which brings the particle of an atom to vibration and holds this most minute solar system of the atom together. **We must assume behind this force the existence of a conscious and intelligent mind.** *This mind is the matrix of all matter."*[7] Max Planck (1858–1947) speaking in 1900

"What we observe as material bodies and forces are nothing but shapes and variations in the structure of space." Erwin Schrödinger

"I sincerely believe that the fabric of reality is composed of a multi-leveled vibrational field that is alive, conscious and intelligent." [8]Hank Wesselman, Ph.D.

"Science is not only compatible with spirituality; it is a profound source of spirituality."
Dr. Carl Sagan (1934–1996)

"I would like to suggest that superstition is very simple: it is merely belief without any evidence." *Dr. Carl Sagan, Gifford Lecture No. 1, 1985*[1]

1. SO WHAT DOES SCIENCE REALLY KNOW?

Not very much, really... As I stated in the preface, recent theories of *Dark Energy* and *Dark Matter*, which have a lot of data to back them up, suggest that we are as yet unable to discern 96 percent of the universe... Which means that what we can discern is only 4 percent of the total. Entirely missing 96 percent of reality, even in our best attempt to construct a standard model of the universe, is itself a monumental failure representing phenomenal hubris.

But frontier scientists now realize, just as the Nobel laureates who gave us quantum electrodynamics (QED) knew as early as 1900, that areas of scientific inquiry which American *Materialists* even today smugly assume to be "impossible" are in fact valid.

Indeed, reductionist *Materialist* scientists suddenly have a great deal of egg on their faces now that the paranormal (consciousness, psi, clairvoyance, remote viewing, extrasensory perception, after-death communication, psychokinesis, precognition, etc.) have all been validated by scientifically controlled triple-blind laboratory experiments and proof of non-locality has been repeatedly replicated.

For centuries, the *Materialist* community has naively, arrogantly, and over-confidently believed that merely because we can measure a thing that we also somehow understand it. Again, this was monumental hubris. To illustrate this lack of understanding, here are several examples of things we just don't know, but that, in Western culture, most people assume we do know and have figured out years ago.

EXAMPLE 1 - GRAVITY ITSELF IS STILL AN UNKNOWN: A good example of this **believing we know everything when we really don't** has to do with gravity. Everyone remembers from a grade school folk tale that Isaac Newton identified the force of gravity one day when an apple fell out of a tree and hit him on the head… But true science is way more difficult than that. A real scientist wants to also know:

- Why does gravity make the apple fall to the earth?
- Why and how does that happen?
- What is this force made of? and
- What causes this force?

Yes, Isaac Newton discovered the existence of the gravitational field, and clever mathematician that he was, he also was able to describe the effects that gravity produced using a mathematical equation he derived that calculated the acceleration of gravity to be thirty-two feet per second squared. Great work, Isaac!

But although he could measure its effects, Newton was never able to form any conception of the nature of the force that caused gravity (the attraction itself). Today, although most non-scientists believe we figured out gravity a long time ago, **our best scientists still know no more about the force of gravity than Newton did 350 years ago.** Instead, Einstein has postulated that gravity is curved and that the mass of planets causes the matrix of space to curve and that is how gravity works.

Unfortunately for Einstein's gravity proof of non-locality in 1972, with its replication in 1981–82, twice in 1998, and then replicated five more times (in 2000, 2001, 2007, 2008, and 2009) leaves no room (excuse the pun) for space to exist. It is obvious that space is entirely an illusion, and that in "reality" there is no space there to curve; space is just a perceptive device, an intellectual dodge that our we use to perceive a physical reality within a universe made of consciousness. So "a curvature in space" is no longer a sufficient explanation.

Consequently, if you have an alternative hypothesis that maybe *the attraction of gravity is caused by objects loving each other*, your hypothesis is actually more advanced than the best *Materialist* science has given us. At least your hypothesis has a motive (love). Whereas the best current *Materialist* "science" knows nothing about the "why" of gravity and it lacks even a motive.

Yet some people, even Ph.D. scientists, naively and smugly act as if **science already knows everything**. But "science" is not a set of hard and fast laws written in stone somewhere, or handed down in some book of holy writ. Science is only a method of reasoning, a way of working out the answers to problems.

EXAMPLE 2 - SPEED OF LIGHT IS NOT A CONSTANT: Another example of *believing we know more than we do* has to do with the speed of light (the "C" that appears in Albert Einstein's famous formula $E=MC^2$). I was taught at the university that this quantity had been proven to be a constant (a never changing value) in physics. This "belief" has remained basic scientific dogma for the last century and a half, ever since Leon Foucault accurately measured the speed of light in 1862. He found that light traveled at precisely 186,000 miles per second. This is still taught in high school and college physics classes all over the world. Einstein's relativity, in order to be true, relies on this being a fixed constant. Without the speed of light being a fixed constant, his formula falls apart. But our cutting-edge physics has demonstrated that this just isn't true.

1. Recently I read an article in *Science* (Feb 20, 2015, Vol 347, Issue 6224, page 856) which cited experimental results proving that "C" (the speed of light) is not a constant, but is a variable quantity depending on several factors[9]. So, I dug a little deeper.

2. Those readers who are practicing scientists and members of the American Association for the Advancement of Science (AAAS.org) can go the AAAS website and click onto "The Cutting Edge" where they can watch a great twenty-two-minute video (from 2011) in which Dr. Lene Hau, Ph.D. Harvard University describes the *Bose-Einstein Condensation and the Taming of Light* and tells exactly how she slows the speed of light (for certain particles) down to fifteen miles per hour.

But more importantly (as will become clear later in this book, after we discuss the Double Slit experiment), Dr. Lau actually cools atoms down to nano-kelvins of temperature, nearly into the zero-point field (ZPF), until they are actually held in limbo, half in the quantum state and half in the physical state. Then, by simply adding a laser light (an observer) she can re-materialize the light. This was published in an article in *Nature* in 2010, "Cycling at the Speed of Light."[10]

If the speed of light can, by simple sub-cooling (and as Dr. Lau has demonstrated), be slowed to less than fifteen miles an hour, then to have believed it was a constant has now been proven wrong and it is maybe time to scrap Einstein's entire theory of relativity; one of his constants (the "C" in E-MC²) is no longer found to be constant.

Again this simply shows the monumental hubris inherent for any scientist ever assuming that they know anything "for sure." Please realize that Dr. Albert Einstein never assumed to know everything. Like Dr. Carl Sagan he knew that *we just don't know.* Einstein instead found ways to explain things to the rest of us.

EXAMPLE 3 - QUANTUM THEORY AND RELATIVITY DON'T EVEN AGREE Finally, we have the disturbing fact that our two greatest scientific theories of the twentieth century do not actually agree. In fact, they each require opposing sets of First Principles:
- **Quantum theory** requires the nature of reality to be *dis-continuous, non-local, and non-causal.*
- **Relativity theory,** on the other hand (based on *Materialism* being true), requires the exact opposite, mandating instead that reality must always be *continuous, local, and causal.*

It is obvious to any thinking person that both theories cannot be true, simultaneously. Therefore, something must be as Alcock said terribly wrong with our perceptions of reality when we actually believe both theories are true... Maybe our problem is in the foundational belief that *there can only be one right answer at a time.*

Today, it is becoming more and more obvious, a hundred years after relativity, that although Dr. Einstein strived mightily to fit the larger universe into classical *Materialist* physics (conceived in the 1600s when we thought there was only one galaxy), any concept of "space" was going to require more "room" in the Newtonian box (again excuse the pun). Quantum physicists simply realize that space is entirely our chosen illusion.

But if we have to choose just one that is true (QED or Relativity), which is most likely? I personally enjoy my cell phone, digital electronics, flat-screen plasma TV, fMRI scans, etc., all of which are based on the reality of quantum electrodynamics being true... But if quantum electrodynamic theory were not true, I'd have to give all these things up as falsehoods. Now after non-locality has been proven multiple times, there is no choice. Relativity and *Materialism* are out and QED with its required conscious universe is in.

Materialists, most of whom could not explain Relativity to begin with, have chosen instead to live with cognitive dissonance, avoiding this choice for eighty years by ignoring it. This dissonance has caused two opposing viewpoints to evolve in how scientists perceive the universe. Either they honestly choose to see things as agreeing with the theory of QED or they religiously choose to see them as agreeing with nineteenth-century materialism and Einstein's Relativity instead.

Obviously to reach the underlying truth we must reconcile this discrepancy before we can reach any level of certainty.

2. RECONCILING THE DISAGREEMENT:

Dr. DAVID BOHM: While working at Princeton alongside Dr. Albert Einstein, Dr. David Bohm decided to construct a theory that could encompass all the known data, and also agree with both QED and Relativity. Bohm had earlier attended Cal Berkeley as one of J. Robert Oppenheimer's doctoral students (with Oppenheimer having been trained by the "English and German Geniuses" (Ernest Rutherford at Cambridge and Max Born at Göttingen), Oppenheimer was friends and fellow student with Heisenberg, Jordan, Pauli, Dirac, Fermi and Teller). Dr. Bohn now worked alongside Einstein at Princeton University and he decided to resolve the disagreement. These two physicists (Bohm and Einstein) began to hold regular discussions about this very inconsistency.

Einstein had previously read Bohm's descriptions of quantum theory and believed that Bohm had given the best explanation that Einstein had seen on this subject. Now, the two scientists set out to resolve this inconsistency and find agreement.

First, Bohm decided to look at where the theories agreed. He found that both quantum theory and Relativity agreed that
> *The cosmos is a single unbroken wholeness in flowing movement.*

He next noticed a hidden underlying problem that there were two aspects of the universe, each of which could only be perceived in different ways. He called them the *explicate* and *implicate* orders, and said that they only appear distinct from each other due to our chosen perceptual limitations.

Bohm understood that the ultimate nature of physical reality is not a collection of separate objects (as it appears to our five senses). Instead, reality is an undivided whole which is in perpetual and dynamic flux. (i.e., it may move in and out of physicality manifesting from the void and annihilating back into the void as needed).

The insights of quantum electrodynamics (QED) and Relativity theory both pointed to a universe that is undivided and in which all parts **merge and unite in one totality.**

This undivided whole is not static, but is in a constant state of flow and change, a kind of invisible ether (*dark energy* or *zero-point field*) from which all things arise and into which all things eventually dissolve. Later scientists studying the vacuum (the ether) discovered that this manifestation and dissolution takes place every ten yoctoseconds, that is twenty-three septillion times a second, such that we, and all matter, are continually reappearing apparitions. It is our rate of vibration which decides whether or not we happen to be visible within the small range of frequencies detected by the human retina.

Indeed, in this undivided universe, even mind and matter are united:
> *"In this flow, mind and matter are not separate substances. Rather they are different aspects of one whole and unbroken movement."*

Similarly, living and nonliving entities are not separate. As Bohm puts it,
> *"The ability of form to be active is the most characteristic feature of mind, and we have something that is mind like already with the electron."*

Thus, matter does not exist independently from so-called empty space; matter and space are each part of the wholeness.

Bohm found that with our five senses we can perceive only the *explicit order* (matter and our physical reality). This is an unfortunate perception that lends itself to reductionist *Materialist* thinking, and causes us to completely miss the larger reality.

On the other hand, everything else that we can't perceive directly through our five senses is the *implicit order,* which is only perceivable through intuition (our sixth sense). This learning to rely on this sixth sense and not the five senses is what I study scientifically in mediumship.

Unfortunately, since childhood most of us have been taught to believe that only what our five senses perceive (*the explicit order*) constitutes our physical reality, and we have also been taught to ignore our intuition (*the implicit order*).

But it was obvious to Dr. Bohm that both orders were actually fully existent, so he worked out the complicated mathematics showing that both the explicate order and the implicate order are fully consistent with the data of QED and also with Relativity, and later that they also do both agree with John Bell's non-locality.

So, we don't have to actually choose. Relativity is a way of perceiving our physical reality, just as Newton's law of energy conservation is also a way to perceive our physical reality, despite the fact that the energy is transferring back and forth from the zero-point field twenty-three septillion times a second. But unless I can use that law of energy conservation, I cannot design a steam plant. On the other hand, merely because Newton's law of energy conservation works handily for me in practical engineering does not mean I should limit my perception to the boundaries of its limited range.

For many scientists, accustomed to discounting the paranormal, it may be difficult at first to see that the fundamental reality is not what appears to our five senses. For some it may also be tempting to assume that the implicate order only refers to a subtle level of reality that is therefore small, "secondary" and subordinate to the explicit order of our primary physical reality, the material substance all around us.

Yet according to Dr. Bohm, the truth is the exact opposite. To Dr. Bohm, the invisible, implicate order is the "primary" reality, even though it is invisible, and the explicate order (the physical universe we experience as visible) is in fact a very crude "secondary" reality. In truth, the physical universe we experience is no more than mere froth on the surface of the ocean of the implicate order.

Bohm's implicate order is synonymous with what most of us call universal consciousness, the universal field, or the Spirit realm, and it is everywhere present just beyond space and time. We cannot directly perceive it except through intuition, meditation, and psychic mediumship. Dr. Bohm also found that our classical science had mistakenly overlooked the implicate order entirely, which is how they missed 96 percent.

To envision this, I like to think of the implicate order as the coffee in my Starbuck's vente and the explicate order as the foam they put on top. The coffee is what I really came for, even though the foam is all that is visible when they hand me the cup. I know the foam is just window dressing and the real caffeine lies invisible just underneath. I love the foam, but it is not the reality of why I drink coffee.

AN ONTOLOGICAL INTERPRETATION OF QUANTUM THEORY: Dr. Bohm saw that the basic nature of reality involves three fundamental concepts (matter, energy, and consciousness). But that up until now, science has concerned itself only with two of those concepts (matter and energy) while ignoring the third (consciousness) entirely.

Also, all of our scientific instruments can only measure the explicate order (matter and energy). Unfortunately, our sensors (telescopes, microscopes, ultrasound sensors) only improve our five senses, but they do not sharpen our sixth sense, which is what must be used to sense and measure the implicate order. Again, the reason I scientifically study mediumship.

3. THE RESULTING DICHOTOMY

In the twenty-first century, we find a society with two different ways of perceiving our reality: *Materialism* and what, for lack of a better term, I will call *Idealism*. The philosophical distinction between materialism and idealism appears to be a question of *whether matter is rather crude and mechanical or whether it gets more and more subtle and becomes indistinguishable from what people have called mind.*

Here are a couple of helpful definitions:

[lb]   **Materialists:** These are people (including scientists) who rely only on their five senses, and so see only the explicate order made of matter in a physical universe, shallowly denying that anything not visible can actually exist. They believe all psi phenomena within the implicate order are "impossible." Usually these are people who do not like living with uncertainty or mystery (i.e., they like to feel secure believing they know everything there is to know).

[lb]   **Idealists:** These are people (including scientists) who perceive the magnitude of what we don't yet know, and frontier scientists who are mostly idealists create the cutting edge of science. They tend to practice on the frontiers perceiving the implicate order through their sixth sense, defining consciousness as the matrix of all reality, and contemplate deeply, thereby realizing that there may be unknown or hidden order in what appears to be random.

THE MATERIALIST'S PROBLEM: If a person does not believe in the implicate order, their mind is simply made up in advance to only perceive the explicate order... Anything beyond their five senses is, for them, "simply impossible." Thus, they are automatically blinded to the truth until such motivation is given them to change their mind. This is the problem for anyone who insists on observing everything only with their conscious mind and so refuses to recognize anything that they perceive with only their hearts.

CONSCIOUSNESS EXISTS OUTSIDE THE BODY: On the other hand, for those with the mental inclination to perceive the implicate order, it has become increasingly evident through modern scientific inquiry that the consciousness dwells outside the body.

For example: Hank Wesselman, a Cal Berkeley grad in anthropology who has spent his career working with indigenous tribes in Africa and Polynesia, says, *"Consciousness is the etheric field"* (dark energy) in which everything is interconnected, and that this is *"clearly articulated by the indigenous tribal peoples at one end of the human continuum and also by the quantum physicists and Zen Buddhists at the other."*

Before the discussion of QED, you might want to digress from the text and watch an October 2015 interview in which I discuss all of this. It is available on YouTube at this URL:

"ON THE NATURE OF CONSCIOUSNESS & SURVIVAL" Interview with Dr. Alan Ross Hugenot
(October 2015) 55 minutes.
https://www.youtube.com/watch?v=yByEQfaD314

4. QUANTUM ELECTRODYNAMICS (QED)
The theory often improperly called quantum mechanics (QM) is actually not a mechanical theory at all. Indeed, naming it "mechanics" tends to reinforce the *Materialists'* nineteenth-century "clockwork mechanism" view of reality.

Instead, it is truly an electrodynamic "hypothesis" and so is more properly named *quantum electrodynamics* (QED), which is the terminology used in this book, by its founders, and the Nobel laureates who finalized the mathematics for it. This is the theory we are most interested in, because it allows for the consciousness to exist outside of material matter, and also gives us our modern digital electronics.

The background matrix, or zero-point field (ZPF) of the implicate order is the ground upon which the universe is constructed, and which is some form of field, similar to an electromagnetic field, but that we cannot as yet discern. It appears to be either within the dark matter and dark energy, or it is a matrix of undiscerned plasma (a fourth form of matter); earlier scientists called this the luminiferous-aether.

Quantum electrodynamics can be simply described as:
Fields, and the interaction of fields…which is all that really exists.

Albert Einstein, although he wanted to continue to believe in the *Materialist* perspective of classical Newtonian physics, which he had adopted when he first learned physics, actually understood that QED fully falsified his beloved *Materialist* paradigm, and so he himself said,

> **"There is no place in this new kind of Physics (quantum electrody-namics) for the field and (also) matter, for <u>the field is the only real-ity.</u>"** *Albert Einstein*

So, materialists don't even have the arch Materialist Einstein on their side. Max Planck, who brought Einstein to Berlin as a professor, improved on Einstein's thought by saying,

> **"As a man who has devoted his whole life to the most clear-headed science, to the study of <u>matter</u>, I can tell you as a result of my research about atoms this much: 'There is no matter as such.'"** *Max Planck (1858–1947), Father of Quantum Mechanics*

5. THE HARD PROBLEM

This brings us to the current crux of scientific inquiry, literally the problem of the ages, which Shakespeare so often illustrated in his plays. *Is there order or chaos in the universe?* Logically, it cannot be both ways.

If (as the *Materialists* hold) our universe is all a "miraculous accident" (i.e., no God), then the universe is ruled entirely by *chaos*. But *Materialists* also believe in *determinism* as an integral part of their paradigm. Now, if all is actually chaos, then it is simply not logically possible for anything to be *deterministic*. This is just another example of the massive cognitive dissonance that is so prevalent within twenty-first-century *Materialist* dogmas. Trying to prop up 350-year-old religious dogmas to account for all the anomalies discovered since merely stretches the dogmas until they come tumbling down like a house of cards.

Yet again, conveniently employing juvenile double-think, *Materialists* naively choose to believe simultaneously in the mutually exclusive concepts of *determinism* and *chaos*. Interestingly, when *Materialists* need order and some policeman to enforce that order, as in *what force makes the laws of physics inviolable,* they conveniently use pseudonyms like "mother nature" or "the laws of nature" when what they really mean is "God (the enforcer)."

On the other hand, if (as the idealists and biocentrists hold) there is order in the Universe, then that order is maintained by some form of universal consciousness (God). Also, if consciousness exists outside time and space, then such a universal consciousness is also a rational conclusion.

The hard problem has been eloquently defined by Dr. David Chalmers, the distinguished professor of philosophy and director of the Centre for Consciousness at the Australian National University, and also visiting professor of philosophy at New York University, where he directs the consciousness project. Dr. Chalmers specializes in the philosophy of mind and cognitive science.

Dr. Chalmers has named the disturbing fact that *consciousness seems to be outside time and space* as "The Hard Problem."

"It is natural to hope that there will be a materialist solution to the hard problem and a reductive explanation of consciousness, just as there have been reductive explanations of many other phenomena in many other domains. But consciousness seems to resist materialist explanation in a way that other phenomena do not." [11]

David Chalmers, P. 4, *Consciousness and Its Place in Nature*

Here is an edited version of Dr. Chalmers' description of **THE HARD PROBLEM**:

1. **How does consciousness arise from unconscious matter?** Materialist science says that matter is inert, dead, unconscious. So, if that is in fact true, then how on earth could consciousness (thoughts) arise from this dead unconscious matter? Obviously not doable.

2. **We are stuck in the wrong paradigm.** Our paradigm is our deep unconscious perceptions within which we do our work.

3. **When we learn new things we must change our paradigm.** This sort of change is extremely difficult because our paradigm is how we understand and explain our reality. Here are the four states of what happens:
   The first paradigm encounters an anomaly.
   Scientists try to explain it within the existing paradigm.
   We call this epi-cycling
   Then a new paradigm comes along and is at first ridiculed by the scientific establishment.
   Then finally it is accepted and suddenly "everyone knows that" and no one will admit the hubris that they previously did not believe it and also actively fought against it.

4. **We are currently going through a metaparadigm shift.** A metaparadigm is the larger worldview that underlies everything. Our current metaparadigm is that the real world is the material world (materialism). But there are several anomalies within this worldview that have not yet reached the level of acceptance to become *real anomalies*. These anomalies are near-death experiences (NDE), reincarnation (children who remember past lives), and after-death communication (mediumship).

11

5. **Yet there is an even larger anomaly, which is consciousness itself.**
   - **Consciousness cannot be doubted**
   - **Consciousness cannot be explained**
6. **The hidden problem is that there are unquestioned** *presumptions* **in the currently accepted metaparadigm.**
   The current unquestioned assumption is the belief that matter is insentient, that matter, our basic material, is totally unconscious. But then how can consciousness have come out of something that is unconscious?
7. **The new alternative metaparadigm our society is struggling to accept is:**
   - **Consciousness is a fundamental** quality of the cosmos, as fundamental as space, time, and matter.
   - But where we need to arrive is at the understanding that consciousness is even **MORE FUNDAMENTAL** than space, time, and matter.
8. **Mind did not suddenly appear**; it couldn't have. IT WAS ALWAYS THERE.
9. **Consciousness has evolved and become more rich**. Consciousness has always been there; it did not suddenly appear. It just became more evolved.
10. **Consciousness is in everything** and everything is conscious.
11. Materialist science **unfortunately never explores the implications** of all this.

The "Hard Problem" is made more difficult because *Materialist* scientists mistakenly believe that the underlying foundational concepts which support materialism (*reality, locality, causality, continuity,* and *determinism*), were already proven as facts when *Materialism* began. But *Materialism* began several hundred years ago, at a time when these First Principles, which appear to disprove consciousness as being fundamental, were merely *presumed to be "self-evident."* In the mid-1600s when Newton and Descartes formed the *Materialist* scientific worldview, there was no way to prove these concepts, and they seemed valid enough (i.e., self-evident). But they remain today **unproven presumptions.** Actually, that would be okay if all of these presumptions had not been completely falsified (disproven) by QED and non-locality. Now they need to be either proven or discarded.

Further, in the last fifty years several prestigious universities, and investigative scientific societies, have sponsored rigorous scientific research in several independent disciplines, which all now point to the reality of consciousness survival in an alternative dimension. This new consciousness survival paradigm correlates findings from research being conducted on:

1. The near-death experience (NDE), (iands.org)
2. Children who remember prior lives (CRPL) University of Virginia, (www.jimbtucker.com)
3. Adult past life memories (APLM), (www.scientificexplo-ration.org/journal/jse_22_3_haraldsson.pdf)
4. After-death communications (ADC), University of Arizona (lach.web.arizona.edu), Institute of Noetic Sciences (noetic.org)

This new worldview of a conscious universe where your mind will survive death is neither religious nor materialist, but is scientifically based on the clear implications of QED and non-locality.

For me this was all proven by my own near-death experience, and verified by my own work as an evidential medium in after-death communications. Consequently, when I say that I *KNOW* consciousness survival is real, it is not a hypothetical theory or religious belief, but comes entirely from personal empirical experience.

6. WHAT HAPPENS AFTER YOU DIE

Aside from religious fears and superstitions, some people worry that death itself will be painful, and others believe it will be more like **the lights suddenly going out and never coming back on.**

Unfortunately, many people raised in the materialist paradigm quite logically believe that **when you're dead you are dead.**

But instead, rigorous replicated scientific research has repeatedly shown scientifically that death would be better described as **lifting a curtain and moving on into the next stage of the continuing existence of your eternal consciousness.**

Which is also precisely what I experienced during my personal NDE. So, I tell it to anyone who will listen. In December 2014, this repeated telling resulted in a thank-you from the other side. I received this message at a Spiritualist church while I was visiting in England. The message came through a medium who did not know me, had never heard of me, and had no idea that I was at the meeting of 125 people. But the message came through with overwhelming evidence that it was from a deceased friend. Although this was not someone I ever expected to hear from, the discarnate friend was publically thanking me for carefully and adamantly telling her, two years earlier, when she was facing impending death from amyotrophic lateral sclerosis (ALS, sometimes called Lou Gehrig's disease): "**Barbara, when you are dead, you won't be dead.**" That simple knowledge (knowing what to expect after death) which I had to tell her when her psychiatrist husband was not listening had saved her so much time on that side after her arrival in the afterlife. She said that she was already helping those coming across who did not know what to expect. She signed off with "I love you and I'll greet you when you get here yourself."

Such demonstrations of its vital importance are why I talk about the "unspeakable" subject of death, all the time.

ERASING DEATH: It is the inclusion of consciousness into the subatomic particle science of QED, coupled with extensive study of the NDE, and scientific study of mediumship that has verified for me and other scientists what mediums have been describing about after-death communications (ADC) since the beginning of writing and literature in 500 BC, well prior to the "old testament" biblical times. All of this also collates well with ancient Sumerian, Egyptian, and Tibetan Buddhist beliefs about repeated lifetimes, namely...

There is no death. There are no dead.

Instead, the collected scientific evidence, when taken altogether, provides replication of the fact that the individual consciousness does in some form survive physical death.

> *"We now have, for the first time in the history of our species, compelling empirical evidence for belief in some form of personal survival after death."*[12]
> Robert Almeder, 1992 Ph.D. professor of philosophy, Georgia State University

From my own NDE forty-six years ago, and my subsequent career in applied physics as a naval architect and marine engineer, as well as a consciousness researcher and evidential medium, I personally know, from both scientific and experiential perspectives, that *the continued survival of our consciousness is absolutely real.*

CHAPTER 2

HOW TO USE THIS BOOK

"There is a new worldview emerging which is based neither on traditional religion or Newtonian physics... There is a shift in authority from external to 'inner knowing.'... It has basically turned away from the older scientific view that ultimate reality is 'fundamental particles,' and trusts perceptions of the wholeness and spiritual aspect of organisms, ecosystems, Gaia and Cosmos. It amounts to a reconciliation of scientific inquiry with the 'perennial wisdom' at the core of the world's spiritual traditions. It continues to involve a confidence in scientific inquiry, but an inquiry whose metaphysical base has shifted from the reductionist, objectivist, positivist base of 19th- and 20th-century science to a more holistic and transcendental metaphysical foundation."[13]
    Willis W. Harman of the Institute of Noetic Sciences (noetic.org)

The purpose of this book is to explain, in the simplest terms possible, that the *materialist* perspective we have been taught is false and fully disproven by quantum electrodynamics (QED), and that the *idealist* perspective of a conscious universe is supported by QED, non-locality, and the latest scientific research. Here is how to comprehend this liberating understanding of reality.

FIRST, CONSIDER HOW WE ALL CHOOSE THE "BELIEFS" WE THINK ARE TRUE: Chapter 3 explains the logic of belief formation. This is essential background for understanding the rest of the book, because the resulting worldview this book will give you is revolutionary, and you will be hard-pressed to explain it to your family and friends. Just as I was after my NDE.

Unfortunately, while most of us (including your friends and relatives) consider ourselves to be "well informed" and therefore holding fully rational opinions on all important subjects, the truth is that most of us have been rather unscientific in choosing our beliefs. We often have simply adopted beliefs from friends or family because we wanted to fit in, and have also been

15

too busy to take any time to think through many important subjects. So instead, our opinions are based on superficial sound bites we accidentally heard on the news, or from friends, and in a desire to appear "politically correct" we choose not to argue with the prevailing viewpoints.

SECOND, DISCOVER HOW MATERIALISM, ALTHOUGH SEEMINGLY SELF-EVIDENT, IS NOT SUPPORTED BY SCIENCE AT ALL: Chapter 4 explains how the widely accepted supposition of a material reality existing "out there" is based on several foundational "truths" originally (in the 1600s) presumed to be "self-evident" in the unscientific culture of the times, and therefore requiring no proof or, at least, were not then provable. But later (by 1934) all of these foundational "truths" (presuppositions inherited from the culture of the 1600s) were found to be completely false by the modern science of QED. Consequently, our modern science shows that the perceived "material reality" is entirely an illusion.

THIRD, REALIZE THAT MODERN SCIENCE FULLY SUPPORTS CONSCIOUSNESS SURVIVAL: In February 2014, leading scientists got together and wrote out a *Post-Materialist Science Manifesto*, describing a perceived reality that reintegrates consciousness and spirituality into science, and therewith providing an entirely new ontology for the twenty-first century. Here is what they feel is true:

> **A Manifesto for a Post-Materialist Science:**
> a) **Mind represents an aspect of reality, which is as primordial as the physical world. Mind is fundamental in the universe** (*i.e., it cannot be derived from matter and reduced to anything more basic*).
> b) **There is a deep interconnectedness between mind and the physical world.**
> c) **Mind (will/intention) can influence the state of the physical world, and operate in a nonlocal or extended fashion**, (*i.e., it is not confined to specific points in space, such as brains and bodies, nor to specific points in time, such as the present*). **Since the mind may non-locally influence the physical world, the intentions, emotions, and desires of an experimenter may not be completely isolated from experimental outcomes, even in controlled and blinded experimental designs.**
> d) *Minds are apparently unbounded, and may unite in ways suggesting a unitary, One Mind that includes all individual, single minds.*
> e) **NDEs (near-death experiences) in cardiac arrest suggest that the brain acts as a transceiver of mental activity,** (*i.e., the mind can work through the brain, but is not produced by it.*

NDEs occurring in cardiac arrest, coupled with evidence from research mediums, further suggest the survival of consciousness following bodily death, and the existence of other levels of reality that are non-physical.)

f) Scientists should not be afraid to investigate spirituality and spiritual experiences since they represent a central aspect of human existence.

The Manifesto for Post-Materialist Science is available at (http://www.explorejournal.com/article/S1550-8307%2814%2900116-5/fulltext) and the website featuring consciousness and post-materialist science (see www.opensciences.org).

Consequently, this book looks at the actual science which supports the perspective outlined in this manifesto, including references to all the publications, experiments, and data that prove all of this. By studying this evidence, **you can arrive at the scientific and non-religious conclusion of** *knowing* **that the afterlife is very real**.

FOURTH: DECIDE HOW MUCH OF THE BOOK TO READ:

1. If you are already a scholar on subjects like the near-death experience and after-death communications (mediumship), as many of my readers are, then you should read Chapters 1-4 and 11-13, carefully reading the table of contents to see if there are areas you are unfamiliar with which were covered in chapters 5-10, for example hypothermic cardiac arrest (Chapter 5). Of course, for the small additional time investment, even on those subjects that you already know, you can get my take by reading the whole book.

2. On the other hand, if you are unfamiliar with any of the subjects in chapters 5-10, then they are discussed here for you to investigate.

FIFTH: GARNER THE TAKEAWAY IDEAS: The important thing is to take home these facts:

1. **Materialism is no longer an acceptable scientific viewpoint.** Instead it is junk science, because it has been disproven by the modern science of QED and non-locality. To believe in *Materialism* any longer is merely superstition. Anyone who purports that *Materialism* is still considered valid science is a pseudoscientist, ignorant, and ninety years behind the times.

2. **Consciousness is the matrix of reality.** QED and non-locality (the latest modern science) provide the scientific basis for biocentrism and a conscious universe.

3. **Biocentrism also allows for a conscious universe and consciousness survival,** and

4. **QED (the most proven "theory" in the history of science) actually requires consciousness to exist outside of physics** (i.e., out of body...a conscious universe).

The details in the book are provided so you can know your belief in an afterlife is scientific. Indeed, in America 72 percent of people over fifty believe in some form of afterlife. It is only a very small but vocal group of pseudoscientific skeptics who don't. So you can now win the discussion with any skeptic (pseudoscientist) *Materialist.* Ask them to put up or shut up and put them in their place. Don't expect to convert them (it's their religion), but you now have the ammunition to win any argument. The facts, and all the scientific data, are on your side.

The discussion in each of the following chapters is as follows:

CHAPTER 3: THE LOGIC OF SCIENCE AND PHILOSOPHY: What Constitutes Truth and How Belief Systems Come About. How we form our beliefs, what constitutes a mathematical proof, a scientific proof, a legal proof, and a philosophical proof (closer to a wish). This background in philosophy of belief and how we culturally prove just "what is true" is invaluable in understanding the rest of the book. This insight allows you to see through controversies by realizing underlying differences in philosophical values.

CHAPTER 4: THREE PERSPECTIVES: Materialism, Quantum Electrodynamics, and Biocentrism. This chapter describes where each perspective is coming from, and also explains why they have their differences. It gives you the tools to see through the misperceptions on both sides.

CHAPTER 5: SCIENCE OF HYPOTHERMIC CARDIAC ARREST (HCA). For most people this will be unheard of, and yet thirty years of medical practice have proven this therapy, where people undergoing brain or heart surgery have been literally placed in "cold storage" without any blood circulation or respiration for three or four hours during surgery, and then are brought "back to life." These patients are, by all definitions, completely "dead" for several hours. Yet many of them have clear memories of what the doctors were doing, which they viewed from the top of the room.

Now, if the consciousness is a byproduct of the living brain (as the materialists want to believe), when the brain is not functioning, there simply can be no consciousness. But here during HCA procedures, the consciousness is surviving somewhere else without the brain or heart functioning at all.

This is also corroborated by the work of Dr. Pim van Lommel, a cardiac surgeon who has scientifically studied cardiac arrest patients whose hearts and brains were also dead by all legal definitions, before they were later revived, but who have conscious memories during the time they were officially dead.

If those consciousnesses (mind or spirit) were having experiences somewhere while their heart and brain were dead by all our definitions of dead, then that must be, by definition, an afterlife.

To prove my point at the end of the chapter, there is a YouTube play listing or eighty-seven separate videos on this emerging therapy being implemented throughout the US.

CHAPTER 6: SCIENCE OF THE NEAR-DEATH EXPERIENCE (NDE). This tells both the story of my own NDE, followed by a full explanation of the proof of consciousness survival which derives from the science of NDEs. Today, most people have heard of the NDE, but you still might like to know my personal story, so it is here. Unfortunately, there are still medical doctors, trained in Materialism, who refuse to believe this is any more than a delusion. On the other side of the street, several Fundamentalist church bodies voted in 2015 to not allow NDE books in their Christian bookstores. Apparently the scientifically proven afterlife is not exactly the way they want it to be. Unfortunately, the scientific afterlife does not include punishment and reward, but does like the St. Paul said in the Bible and treats everyone the same. *"Heaven is a free gift, it is not earned or deserved"* (Ephesians 2:8-9). But voluntarily putting on the same blinders as the *Materialists* won't make this disagreeable anomaly go away for their religions any more than it did for the *Materialist* religion.

CHAPTER 7: SCIENCE OF CHILDREN WHO REMEMBER PAST LIVES (CRPL). Dr. Ian Stevenson, Dr. Bruce Greyson, and now Dr. Jim Tucker have been successive leaders for over fifty years at University of Virginia, investigating children who remember past lives. In over 2,500 cases, the details children remember are investigated on location to see if the facts can be corroborated. Amazing details are "remembered" by these children about their prior lives.

Although the single-life hypothesis is the principal belief in Western cultures, the majority opinion in the world is that we live multiple lifetimes. Only the Roman Empire and its descendants remain steadfast in this one-life hypothesis. Origen, the Christian bishop of Alexandria in 200AD, spoke about Christian reincarnation. This belief in reincarnation was outlawed by Emperor Justinian (not the Pope) for the entire Roman world in 553AD. Indeed, there is appearance in the official records and minutes of that council where this alleged "anathema" against Origen was passed. Many historians today believe Justinian made it all up, just so he could use the church to impose his will upon society.

CHAPTER 8: SCIENCE OF AFTER-DEATH COMMUNICATIONS (ADC). Here, I tell my own story of scientifically investigating mediumship. I have personally made an extensive (ten-year) scientific study of mediumship and find it is absolutely valid. Today, I regularly provide mediumship readings for complete strangers and am often able to bring through undeniable evidence of departed loved ones which I could never have guessed. Frankly, where could this information be coming from other than the surviving consciousness of departed loved ones?

Since 2014, I have been repeatedly tested as a research medium under double-blind and triple-blind conditions, and even on occasion with simultaneous, thirty-two-point EEG studies of my brain while doing mediumship. Working with all of the following laboratories:
1. Consciousness Research Lab (CRL) at the Institute of Noetic Sciences (noetic.org), working with Dr. Dean Radin, Ph.D. and Dr. Arnaud Delorme, Ph.D. in 2014 and 2015, and I continue to work with them and the CRL.
2. Laboratory for Advances in Consciousness and Health (LACH) at the University of Arizona (Lach.web.arizona.edu). Since 2015, I have been doing extensive work with Dr. Gary Schwartz, Ph.D., director of LACH, and will continue for the foreseeable future.
3. Psychical Research Foundation (psychicalresearchfoundation.com). In January 2016, I began working as a research medium with Bryan Williams, assistant director of research at PRF.

The research at IONS and at LACH shows that other test mediums and I are not imagining the evidential details we receive; indeed the information has been shown to be coming from outside the mediums. Currently the most logical scientific explanation (our best science) is that these evidential details are actually coming from the surviving consciousness of the departed loved ones. Of course, this is just as mediums have been saying for over 2,500 years.

CHAPTER 9: SCIENCE OF PAST-LIFE REGRESSION HYPNOTHERAPY (PLR). Although you may know someone who claims to have been Cleopatra or Napoleon in a past life, there is also a lot of very credible data coming through people who undergo past-life regression hypnotherapy. However, *we only need one white crow to prove that all crows are not black,* and there is some excellent data coming from honest scientific inquiry, which confirms survival.

CHAPTER 10: SCIENCE OF REMOTE VIEWING. The U.S. Government funded the Stargate project at Stanford Research Institute for over twenty-four years. Stargate thoroughly investigated "clairvoyance at a distance," the results of which the military intelligence community utilizes to find out what is going on where recognizance photography is ineffective. Today there is a U.S. Army protocol for remote viewing, used by the intelligence-gathering community, which teaches how to go into the altered state required to access the sixth sense for remote viewing. I know personally two of the scientists who worked on the Stargate project, Russel Targ and Dr. Dean Radin.

This area of research seems to prove out-of-body consciousness. Somehow, the remote viewer accessed the information, and for the best remote viewers the information seemed to be coming to them as they walked around at the remote location. And if one thinks of our universe in terms of quantum fields and QED instead of material objects, suddenly this entire concept makes sense. The mind may not necessarily "travel" spatially to this remote location. Indeed, within our non-local universe "travel" may be merely an illusion. Instead, the mind might simply *access* the information, which may be stored holographically throughout the universe. This will all require additional research to sort out.

CHAPTER 11: SCIENCE OF QUANTUM ELECTRODYNAMICS (QED). It is a little-known fact that quantum electrodynamics requires a living consciousness to observe something in order for the matter of our physical universe to form. Here, we will look closely at the double slit experiment to see how even mechanical observations taken by an electromechanical sensor will remain in the quantum state until some living consciousness actually observes the recorded data. In other words, the cat is not dead or alive until someone opens Schrödinger's box. Instead, it remains in the undetermined "quantum state" until it is observed by a living consciousness. We will also look at how John Bell's theorem of non-locality supports and agrees with all this. We will look closely at the Von Neumann interpretation of QED, as well as dark matter and dark energy.

CHAPTER 12: SCIENCE OF BIOCENTRISM. Here we will be talking about the science of consciousness outside the body, which adds fuel to the fire that consciousness existed before matter, and so is a basic constituent of matter. Biocentrism is a new view of science that places physics at the side and instead sees that "life" or "consciousness" was here before the Big Bang. It is a view that sees consciousness (life) as the basic building block of the universe.

CHAPTER 13: CONCLUSION - WHAT ALL THIS MEANS. Here we sum up all of the things we have examined and what the implications of all this actually are in light of a conscious universe and consciousness survival in an afterlife.

APPENDIX A: THE WEIGHING OF THE HEART *"La Pesée du Coeur."* This is an excerpt from my previous book, which concisely examines and explains the history of religious traditions since the ancient Egyptians. It is a little uncomfortable for fundamentalist Christians and Muslims, because it shows that all religions are descendants of this single ancient religion, which is the sort of exposé of history that makes them want to "kill the infidel" (destroy the messenger), but it is the honest history of religious superstition since the dawn of time.

APPENDIX B: SPIRITUALISM & QUANTUM ELECTRODYNAMICS: A COMING PARADIGM SHIFT. This looks at the melding of Spiritualism and science. This is literally *the Philosopher's Stone* which the alchemists sought since the Middle Ages in order to reunite Western science and spirituality, which began to bifurcate in the 1400s as science split from the church, and was finally stated as the dualism created by Descartes. This will reintegrate the Western soul as the descendants of the Roman Empire reembrace the eternity of the individual spirit, an obvious concept that the Eastern soul (non-Roman Empire descendants) has never lost.

CHAPTER 3:

THE LOGIC of SCIENCE and PHILOSOPHY, WHAT CONSTITUTES PROOF and HOW BELIEF SYSTEMS COME ABOUT

"A new scientific truth does not triumph by convincing its opponents and making them see the light, but rather because its opponents eventually die, and a new generation grows up that is familiar with it."[14] Max Planck, Scientific Autobiography and Other Papers

"Pseudo-skeptics often are typically disbelievers — i.e., they are firmly entrenched in believing 'no' about certain things. Although they may 'claim' that they are open to new information, they typically react with strongly unfriendly if not hostile criticisms when their beliefs and assumptions are challenged by new ideas and evidence... Pseudo-skeptics typically make extreme statements. They will sometimes categorically state that something is impossible, or they will make sweeping false statements such as 'no evidence exists,' or the experiments are 'all flawed' or even that the scientists in question are engaged in 'pseudoscience.'" —Dr. Gary Schwartz, Ph.D.

Clarity about our beliefs, and those of others, can easily be found by simply shortening the phrase "Belief System" down to two letters: BS... This insightful exercise places all beliefs (yours and everyone else's) on an equal footing...just everybody's personal BS.

Unfortunately, while we each think our own *beliefs* are actually *facts*, and so can't understand why someone else does not agree with these *facts*, the sad truth is they all are just beliefs (merely things we hope are true) and as such they are instead just belief systems (BS). Realizing that all beliefs are built on speculation and not facts, this convenient shorthand of referring to all as BS reminds us that any belief is flimsy, and even our personal chosen BS is just as flimsy and unsupported as the next person's BS.

Looking at the world this way helps us not to fall into the psychological trap of fundamentalism and desiring to **"kill the infidel."**

Consequently, throughout this book I will refer to all belief systems with the initials BS.

Another insight I find helpful is remembering never to argue for your cause, because as Buckminster Fuller said,
> *"You never change things by fighting the existing reality… To change something build a new model that makes the existing model obsolete."*
> —Buckminster Fuller

But in order to understand why many "supposedly scientific" people have such difficulty with the concept of consciousness survival, or even consciousness in general, it is important to understand the several opposing systems of logic that define "truth" in various intellectual fields; hence, the following discussion is essential to any cooperation with people of differing intellectual disciplines.

1. WHY DO WE BELIEVE WHAT WE DO?

How can we ever be positive that what we know "for sure" is actually true? When we examine our own "known" values, we may discover that we believe something a certain way only because, as Grace Hopper (U.S. Navy admiral in computer science) said, "We've been following the most dangerous words in the English language: '*We've always done it that way.*'" All too often these "ways" are handed down for several generations without looking at "why" we do it that way.

> For example: *The story is often told of a woman who always cut off the ends of the roast before she put it all in the roasting pan. One day her husband asked why she did that… "Because my mom taught me to do that. It's what you do before you put it in the oven," she said.*
>
> *Later that year, she asked her mom why they did that. Her mom, who was in her late sixties, said, "Because Grandma taught me to do that; we've always done it that way."*
>
> *But at Christmas, they both confronted Grandma, who was in her late eighties, and asked her why she always cut off the roast on both ends. Grandma looked at them, a little confused, and said, "Because the roasting pan was always too small for the roast. Why do you ask?"*

It is good to remember that at least some (if not all) of our own belief systems (our BS) are based on the same kinds of presumed evidence, which was never thought through.

24

Admiral Hopper also taught one other Navy axiom that I use every day: *"It's easier to ask forgiveness than it is to get permission."*

2. HOW REAL SCIENCE WORKS:

The ideal in science is to allow our experiences, in the form of formal observations and measurements, to rationally shape our understanding and therefore our beliefs. Normally we use the two-step process we call *the scientific method*:

1. We carry out controlled experiments hoping to find data that supports our prior hypothesis. But
2. If we find data that does <u>not</u> support our prior hypothesis, then we must revise the hypothesis to fit the new data.

Not too complicated really, and it will always result in the truth if it is actually followed out.

Unfortunately, if we are a *Materialist* scientist who has spent thirty years investigating an area of science, when new parapsychological data comes along that completely destroys what we have given our life's work to, we would not be too eager to embrace that new data. So we would then ignore the new data. This is why Max Planck said that "science only advances one funeral at a time."

Yet if as a scientist we violate the scientific method by ignoring disagreeable data or not taking into account all the data, then the scientific process breaks down entirely and ceases to be science and becomes instead pseudo-science.

This is where the *Materialists* who are the loudest skeptics, opposed to all things "paranormal," continually err. They give loud lip service to the scientific method,", praising it and declaiming loudly that *"science has proven that the paranormal is impossible."* They then proceed to ignore the data proving the paranormal, and with no data of their own with which to disprove it, they must then proceed to vilify the actual scientists who are honestly researching the paranormal. Calling real scientists frauds is not the scientific method; it is just stupid and dishonorable. Such shallow charlatans are not scientists but are most definitely pseudoscientists and religious fools.

In order to maintain strict scientific conditions, a true scientist must always revise their hypothesis to suit ALL THE DATA, regardless of how uncomfortable they are with that data or the revisions it forces in their pet hypothesis.

Prior to the discovery of quantum electrodynamics (QED) in 1934 and its mathematical formalization in 1965, most scientists believed that the universe was like a great big clockworks, with everything happening according to the laws of physics (as defined in their First Principles). These laws of physics could not be violated (which in itself seems a bit strange since when *Materialists* hold that there is no God to enforce these inviolable laws). But "Promissory *Materialism*" believes that all we yet need to do in order to "know everything" is to discover all the "missing" laws of physics.

At the end of the nineteenth century (before the quantum revolution) it seemed that *Materialist* science had discovered most of those laws, and that from then on, scientists would just be "mopping up" one or two unexplained anomalies. The standard model was based on the **First Principles** of *Materialism*, which left little room for doubt, and according to those First Principles, some things were simply impossible, and there was certainly no sense in wasting effort investigating things that were already known to be impossible. It appeared that they knew almost everything else. Some government officials were so sure of this that among other things, they considered reducing staff at patent offices.

But then in the early twentieth century, Einstein, Planck, Heisenberg, and that gang of "German Geniuses" removed all that comfortable certainty with the advent of QED. Suddenly, the underlying science shifted, and the prior Newtonian materialist theory that everything is made of matter (substance) immediately became obsolete. Suddenly, with QED, *matter is not made of matter*. Instead, everything is made of fields.

Yet for those *Materialist* scientists who were comfortable in a world of solid matter, to stop believing in matter itself would require discarding the comfortable scientific paradigm of *Materialism*, with all its prior speculations and superstitions, and to instead reconstruct all the theories to agree with QED and quantum field theory. One prominent skeptic (who has a bachelor's in physics, but his Ph.D. is in psychology, so he is a psychologist and not a scientist), James Alcock, stated the problem clearly as follows:

> Dr. Alcock said that **the claims of parapsychology (and QED)** *"stand in defiance of the modern scientific worldview."*...And *"if the claims of parapsychology prove to be true, then physics, biology and neuroscience are horribly wrong in some fundamental respects."*

Dr. Alcock is telling two deep truths.
1. The claims of parapsychology do stand in defiance of the *Materialist* worldview, and
2. If we live in a conscious universe (which parapsychology appears to indicate), then *Materialist physics, biology, and neuroscience are horribly wrong in some fundamental respects.*

But Dr. Alcock is also using deceptive syntax to perpetrate several lies, because:

1. The *Materialist* worldview has nothing to do with the modern scientific worldview.
2. *Materialism* is not modern (it is 350 years old), and
3. *Materialism* has been completely debunked, by QED, for over eighty years.

Unfortunately, the claims of parapsychology have proven to be completely true in carefully controlled triple-blind scientific experiments, as supported by QED and non-locality; and just as Dr. Alcock's premonition stated, it is now obvious that something is *"horribly wrong"* with *Materialism* *"in some fundamental respects."*

Indeed, all the fundamental **First Principles** of *Materialism* (which back in 1650 were always mere suppositions, presumed to be self-evident) have been proven to be invalid by QED, as I will explain in the next chapter.

So, just maybe, Dr. Alcock, it is time for you (and your skeptical cohorts) to get on with making the necessary corrections to all these disproven theories instead of wasting all your energy undermining parapsychology and vilifying honest scientists. Instead, do the world a service of finally correcting what is so obviously *"horribly wrong"* with the archaic *Materialist* paradigm. The challenge, Dr. Alcock, is why not be contributive instead of destructive, or as I learned to say in the US Navy,

"Either lead, follow, or get out of the way."

Unfortunately, for many scientists, revising everything they previously thought they "knew" would be nearly catastrophic. So, instead they may defiantly choose to continue to "believe" in *Materialism* even though they know better. After all, *Materialism* is a very good approximation of the physical world, and its formulas work well for most practical applications. So if your lifetime of work is embedded in this broken model, would you gain anything by honestly moving on? It might be the better part of valor, at least as far as your own welfare, to ignore the truth and hold on until retirement.

However, this chosen mind-set quickly forgets that it is based only on unproven BS (*belief system*), and not actually true. And within that *Materialist* BS is a *rigid stupidity* that also excludes many things as "impossible," and this rigid stupidity has dogged science for over a hundred years, as pointed out by William James, who died in 1910:

"I believe there is no source of deception in the investigation of nature which can compare with a fixed belief that certain kinds of phenomena are IMPOSSIBLE."

But to complicate matters further, the particular philosophical discipline a student is trained in determines what their worldview will be and how they will evaluate science. So, let's take a moment to consider philosophy and how it affects our BS (belief systems) about what is true and not true.

3. WHAT CONSTITUTES A SCIENTIFIC FACT?

As a neophyte undergraduate, in the naïve days before I studied both law and philosophy, I did not realize that there are actually four levels of proof in common use within our Western culture. While each is used by different disciplines to establish what is a "fact," there is no agreement between disciplines on what constitutes "proof." When someone says, *"That's a proven fact,"* it could have four different interpretations. Here are those levels of proof:

  A. Proof **"beyond all doubt,"** which is only applicable to mathematical certainties;
  B. Proof **"beyond all reasonable doubt,"** which is the level used by the hard sciences;
  C. Proof based on **"the preponderance of evidence,"** which is what the courts of law use; and
  D. Proof based on **"it is probable,"** which is utilized in the philosophical sciences. But here what is "probable" depends on what is "possible" within that person's underlying BS.

Obviously not all intellectual disciplines are working with the same set of rules for proof of what is a fact. Here is how these levels of proof are qualified and applied:

A. PROOF BEYOND ALL DOUBT (A Mathematical Fact):

Very few things can be proved to the highest level of being **beyond all doubt**. So, this is mostly reserved for mathematical truths. For example, two plus two is four. But if you get two plus two is five, then it is pretty obvious you are either insane, or you work as a well-paid Wall Street CPA trying to fool the Internal Revenue Service. (Of course that is said with tongue-in-cheek.) Yet most things in science cannot be proven to this level of **beyond all doubt**.

Consequently, scientific facts are required to only be proven to the second level, of being **beyond a reasonable doubt.** But all scientific facts are actually only hypotheses and/or theories, and never actually facts that have been proven **beyond all doubt**. Yet the requirements are still pretty stringent for "scientific facts."

B. PROOF BEYOND A REASONABLE DOUBT (A Scientific Fact):

For a fact to be proven **beyond a reasonable doubt**, hard science requires several things in order to "scientifically" prove a fact to that level:

  1. It must be a **logical possibility**, but more importantly:
  2. There must be **evidence for it,** and
  3. There must be **no evidence against it.**

This level of proof is precisely where the hard science of consciousness survival stands today.

- Firstly, **consciousness survival is a logical possibility**.
- Secondly, **there is a great deal of evidence showing consciousness survival** coming from several separate streams of scientific inquiry, all of which collate with the same scenario of consciousness survival in an afterlife.
- Thirdly, although many skeptical *Materialists* loudly claim that consciousness survival has been completely disproven by science, the truth is that no scientist has ever produced one shred of scientific data to disprove consciousness survival. So, **there is no evidence against consciousness survival.** Instead, there is merely a lot of denial based on absolutely no evidence but a lot of *Materialist* superstition.

Consequently, **consciousness survival is now a scientific fact**, proven beyond a reasonable doubt, which qualifies it as a *fact* in the hard sciences.

C. PROOF BY PREPONDERANCE OF EVIDENCE (A Legal Fact):

This obviously lower level of proof is used by the courts because they have to decide what is truth at a time when both sides in the dispute are lying through their teeth. For example: a divorce case might hear two opposing stories:

1. "He stole all the money out of the marriage accounts just so he could run off with that trollop," as opposed to the other side's argument that
2. "No, he did not take all the money. His business failed for several years, which is why his wife sued for divorce, and the so-called trollop is just his faithful secretary who looked up to him and liked him more than his wife did, and so was willing to stand by him as the business failed, where his wife (who originally married him for his money) chose to desert him entirely…"

In the end the judge must decide what is legally "true" even though nothing can be proven to a mathematical certainty (i.e., a mathematical certainty of "beyond all doubt"), or even to "beyond a reasonable doubt." So, the courts have chosen to allow cases to be decided on the **preponderance of evidence**. In other words, the side that piles up the most plausible "evidence" wins. This is a less reliable level of factual proof, because sometimes the preponderance of evidence is just the tallest pile of lies, but that tallest pile is indeed "legal" in most courts worldwide.

Yet here again, because there is evidence for consciousness survival, and absolutely none against it, the preponderance of evidence proves consciousness survival. So, **consciousness survival is also a legal fact.**

D. PROOF THAT IT IS PROBABLE (A Plausible Fact or a Philosophical Fact):

Finally, we have come to the absolute lowest standard of proof for "facts" which derives only in the "soft" sciences, including philosophy, of which psychology itself is a sub-branch. Unfortunately, this level of proof is similar to magic. Here, we come to the **pseudoscience of "what if,"** *and consequently, these so-called sciences use the logic of probabilities rather than proven realities. But since probabilities are only mental constructs or fictions, these rather undisciplined disciplines only require their "proofs" to conform to the following criteria:*

1. It must be **a logical possibility,** and
2. No evidence either way. Yes, that is all that is required. No evidence is required that the possibility has ever been found to exist.

But this muddle-headed system of proof works out to be *"If it **could** be true, then it **must** be true."* Such illogic is nothing more than Alice in Wonderland magic, which allows persons trained in these soft sciences to so adamantly contradict each other. They simply have no need for data or evidence.

Obviously, this last level of proof is far too simple: The "FACT" is left with no requirement for any evidence to be found to substantiate the probability. Instead, so long as the probability, when clearly stated as a logical possibility, does not conflict with itself, then it is assumed by philosophers to be proven as a valid possibility. This magical-thinking methodology of proof allowed in the soft sciences for proving things is an anomaly when compared to the discipline of the first three levels of proof.

However, knowledge of this anomaly is very helpful when listening to the arguments of skeptics. Each of us thinks and analyzes things in the ways most familiar to us, and familiar ways are usually those taught by our mentors. Consequently, it is not surprising that most of the world's materialist skeptics were trained in the soft sciences of philosophy. This is why they have such a difficult time believing the data of the paranormal. Trained to believe they are being scientific when using *un-provable proofs,* their logic says that:

1. **It is a logical possibility that consciousness survival might not be true.**

So, it is now a proven psychological fact that **consciousness survival is <u>not</u> true.**

Yet the very same logic, with no requirement for evidence, can be used to say that:

1. **It is a logical possibility that consciousness survival might be true.**

So, it is now *also* a proven psychological fact that **consciousness survival <u>is</u> true.**

Realizing this, it quickly becomes obvious why philosophers don't get much respect in our twenty-first-century scientific world, and why those trained in this field find no need for any data, and treat evidence so lightly.

4. A DIFFERENCE IN THE LOGIC OF SCIENCE AND PHILOSOPHY:
From the above discussion it is easy to see that there is a great difference in proof required to establish a fact within the different intellectual disciplines of philosophy and hard science.

HARD SCIENCE FACTS: Hard sciences, like physics and chemistry, are concerned only with *provable facts*; they want replicable answers.

SOFT SCIENCE FACTS: On the other hand, soft sciences like philosophy are fascinated with *probable facts*, building entire theories on "What if." So, it was not surprising to note that most *Materialist* skeptics come from philosophy and not the hard sciences, and that most physicists are neither skeptics nor *Materialists;* instead they maintain a wait-and-see attitude.

PSEUDOSCIENCE: Finally, there is *pseudoscience*, which simply **ignores all facts.** Instead, pseudoscientists believe their chosen assumptions to already be proven facts. This level of logic is not science at all, yet it is the reasoning used by most skeptics, regardless of how "scientific" they pretend to be. And unfortunately, it is the level of arrogant logic (or abject stupidity) used by most *Materialists* when they decide, *"I'll just ignore any data of the paranormal since I already know that it is impossible; therefore, any data must be fraudulent, or mistaken, so why waste my precious time looking at it."*

5. THE PROBLEM WITH SKEPTICS:
Unfortunately, along with honest parapsychological science, there have always been a few outside skeptics who loudly decry everything that parapsychological scientists continually prove. The skeptics spend a lot of time suspecting paranormal scientists of being fools, frauds, or charlatans and so most of their writing about paranormal scientists is pure ridicule and character assassination.

SKEPTICS, POLITICIANS, and BS ARTISTS, ALL START WITH A CONCLUSION:
Unfortunately, instead of using science to generate experimental data to find out what is true, skeptics invariably take the obvious shortcut (just like political pundits). They simply:

    A. Choose what they want to believe first;

    B. Look for facts to back up their conclusion; and

    C. Ignore any conflicting data.

This exercise in mental blindness is an attribute of skeptics in both science and religion (whether arguing for or against any concept). It is also a tactic used often by politicians trying to get elected, regardless of party:

Materialist skeptics simply start from their chosen BS (belief system) and then search for facts or pseudo-facts and pseudo-data that back up what they have already decided to believe. Next, they conveniently ignore any facts or data that tend to disprove what they already decided to believe. Finally, these complete frauds accuse the other side of being fraudulent.

SKEPTICS ARE NOT SCIENTISTS: Considering the levels of proof discussed above, it is not so surprising to find that most all the disbelieving, *Materialist* debunkers and skeptics are not scientists. In fact, if they do hold graduate degrees, they have Ph.D.'s in *philosophy,* and I have not found one who holds an actual "scientific" doctorate in science. Consequently, skeptics all come from disciplines that don't require hard scientific facts in order to prove their own BS. Philosophically, *"I don't think so"* qualifies as a logical possibility, so within their own opinions they have all the philosophical "facts" they need to prove their BS to themselves.

But this will never amount to science, nor prove anything to real scientists. In light of this, *there are no truly scientific skeptics, but only philosophical skeptics.*

On the other hand, there are today, and always have been, true scientists who hold actual doctorates in the hard sciences, who have studied the evidence and believe in consciousness survival. Two eminent British scientists from the early twentieth century who were quite candid and outspoken on their belief in the paranormal were Sir Oliver Lodge (who along with Nikola Tesla is the inventor of the radio) and Sir William Crookes (inventor of the cathode ray tube), both knighted members of the Royal Society. Today, we have even more eminent hard science doctorates who hold these views, like former astronaut Dr. Edgar Mitchell, founder of the Institute of Noetic Sciences.

HOW DO SKEPTICS APPEAR TO REAL SCIENTISTS: Honestly, they all look like shallow-thinking buffoons to any real scientist.

Here is what **Dr. Nancy Zingrone,** a true scientist of the Parapsychological Association and also of the Rhine Research Institute, stated in her 2006 doctoral thesis, which was itself a study of the skeptical movement including Doctors Michael Schermer, the late Gordon Stein, and Susan Blackmore. Here is what Dr. Zingrone said:

> *"Can it be true that many critics behave as if they have never noticed how complicated the world really is... As if they have never turned their focus inward? ... When one reads their writing in a systematic way — from*

pages of the **Skeptical Inquirer** (Michael Shermer is publisher) *to the entries in Gordon Stein's* **Encyclopedia of the Paranormal,** *to their various book-length treatments of the paranormal — one gets the impression that what characterizes the genre* (of skeptical debunking) *is EXCEPTIONALLY SUPERFICIAL REASONING.*"

6. THE LOGIC OF PROOF:

I will illustrate how these different levels of proof operate in the following facetious example:

DO I (Alan) HAVE A BROTHER? *First, I currently believe that I am the only living male child my mother ever produced. Of course, I am not certain about this because I cannot prove it to a mathematical certainty. Unfortunately, I have only enough proof for a scientific fact:*

    1.   **It is a logical possibility** *that I am the only son;*
    2.   **There is evidence for it being true** *(i.e., my parents never told me otherwise); and*
    3.   **There is no evidence against it,** *because I have never met another son of my mother, or my dad, nor anyone who claims to be their son.*

So basically, I have a scientific fact, **proven beyond a reasonable doubt,** *that I am the only son. But on a philosophical level, I still can't be sure.*

Philosophically, <u>*I cannot know this as a fact,*</u> *because using soft science criteria of considering all logical possibilities, it still could be that there is another son out there.*

First, my parents could have concealed this truth from me. For example, during the two years when my father was overseas during World War II, my mother could have had an affair with an outsider. When Dad left to go overseas, Mom was pregnant with my sister, born a few months later. But Mom could have then had an affair, had another son, and then given that son up for adoption. She could have done this all without ever telling Dad. My sister, at one year old, would have been too young to notice the boyfriend or the other baby. This isn't likely as my mother was a Fundamentalist (Nazarene-Baptist). But philosophically, I can't rule it out.

Another possibility? The two miscarriages that my mother reportedly had in the late 1940s in the four years between the birth of my oldest sister and my second older sister. Either of these supposed miscarriages could have been live births, quickly given up for adoption because my parents were living near the poverty line, unable to afford another child. They could have told their friends that the pregnancies had ended in miscarriages to cover up the fact that they gave one or both children up for adoption.

*So, philosophically I can never be completely certain until my "missing brother" FINALLY DOES NOT SHOW UP... But just when is FINALLY? Although I can never be certain, **I do know it is a fact, established to a scientific certainty, that I am the only male child.***

*Yet if I were a philosopher trying to debunk my proof of being the only son, rather than a scientist looking for a scientific fact, then I would have to say that my being the only son **hasn't yet been proven**. This is merely because **a logical possibility exists for another brother to possibly have been born.***

*Now, in a court of law, should this unknown brother show up at the reading of the will, holding a valid birth certificate stating that my parents were also his parents, **regardless of any DNA testing, he would also inherit under the present law**. This is because his valid birth certificate would constitute **the preponderance of evidence**. Yet to be certain **beyond a reasonable doubt** (to a scientific certainty), DNA evidence would have to agree as well (i.e., no conflicting evidence). Even if the DNA did not agree with the birth certificate, the court could still decide, based on the **preponderance of evidence**, that the interloper was legally my brother, even though not biologically my brother. The court could hold his birth certificate, which is a legal document, as superseding any genetic science like DNA, and so his birth certificate would have legal precedence, allowing him to inherit equally with me. Naturally, I would sue for a jury trial.*

*However, to a philosopher, there would be no question: I definitely have a brother. After all, it is a logical possibility, isn't it? So, according to the philosopher, I have a brother... That is, unless you asked him instead if I didn't have a brother, in which case he would also say, **"Yes, his not having a brother is also a logical possibility, so it must be true as well."***

*"**What is real?**" said the rabbit. And by the way, that rabbit was stuffed.*

7. CONCLUSIONS:

So now that we understand what qualifies as a scientific fact, let me say this about the skeptics' viewpoint. Although philosophically, consciousness survival <u>not being possible</u> is one *logical possibility*, to date, no scientist has produced any evidence supporting this *logical possibility*. So, there is, to date, no evidence that our consciousness does not survive.

Consequently, without any such proof, **it then becomes a complete lie to say, "Science has proven that the afterlife does not exist."** But that complete lie is what skeptics regularly publish as being a proven fact. Unfortunately, such proclamation of deeply held BS does not qualify as science.

If, instead, skeptics honestly stated, **"Philosophy has found plausible possibilities that may disagree with consciousness survival,"** then they would be telling a truth, and also properly stating it scientifically.

Consequently, the only honest statement for any skeptic to adopt is what the late Dr. Carl Sagan often said: *"We just don't yet know."*

On the other hand, because there is a great deal of scientific evidence proving consciousness survival and none disproving it, **consciousness survival fully qualifies, beyond all reasonable doubt, as a scientific fact.** Proven by the following criteria:

1. It is a logical possibility;
2. It is supported by a great deal of evidence from several streams of scientific inquiry; and
3. There is absolutely no evidence to prove it is not a reality.

Consequently, consciousness survival is already a scientific fact. and you can say this honestly and confidently to any skeptic who thinks otherwise.

CHAPTER 4

THREE PERSPECTIVES - MATERIALISM, QUANTUM ELECTRODYNAMICS, and BIOCENTRISM

Currently, there are three different paradigms used by scientists to interpret the world and make sense of things:

1. *Materialism* (archaic superstition left over from seventeenth-nineteenth centuries, but falsified by QED);
2. Quantum electrodynamics - QED (understood since 1934); and
3. Biocentrism (emerging since the 1990s).

Each of these starting points comprehends consciousness and the afterlife differently, and in ways that may conflict with the others. However, frontier scientists, able to step beyond traditional materialism, are now beginning to collate the last two (QED and biocentrism) into a single cogent perspective on how the universe works, a new metaparadigm.

MATERIALISTS **CHOSE TO EXCLUDE CONSCIOUSNESS FROM THEIR PARADIGM:** When Rene Descartes and Isaac Newton put together the *Materialist* paradigm 350 years ago, they realized that they could not then explain consciousness. So they purposefully set consciousness outside of physics, in a place they called "meta-physics," meaning "beyond physics." Even today, consciousness cannot be explained using the standard model of physics, where it has been purposefully excluded.

DEDUCTIVE REASONING: The way we choose to reason also colors our conclusions. Deductive reasoning is a marvelous tool, which starts from what is known to be true, and then builds upon that foundation with a logical system of additional postulates to reach new conclusions. It is a tool of reasoning given us by Aristotle, and it is wonderful for solving problems, because we can start with a known truth and just add other truths until we reach a new conclusion, which should also be true.

For example: if the grocery store always stays open until 9 p.m. (a known fact) and you add to that the additional true fact that it is only 7 p.m., deductive reasoning allows you to conclude that you have time to get to the store before it closes. However, if your presumption (presumed fact) that the store always stays open until 9 p.m. has been invalidated by some other fact that you were not aware of (maybe a power outage), then your deduction that you can get to the store before it closes may no longer remain true.

This is what has happened to the *Materialist* reality; what once seemed to be a true fact is no longer true. The foundational presumptions (First Principles) upon which the belief in a material reality was built, all appeared to be true enough in 1650 to be considered self-evident for Descartes and Newton. Yet 300 years later, in the early twentieth century, each of these **First Principles** was shown, by the facts of QED, not to be true.

OPEN-MINDED SCIENTIFIC SKEPTICISM: Understanding consciousness requires a shift of perspective to a larger view of reality, which looks beyond the restrictive limitations of *Materialism*, as originally defined by Newtonian physics as completely contained within "3-D plus time." Instead, we need to move out of the box of "physics" into "metaphysics" (beyond physics) to a metaparadigm, which includes both.

To escape the trap of restrictive reductionist thinking (always breaking things down into little pieces), we also need to be open to new experiences and expansive thinking. Comprehending consciousness is an *aha* experience, which requires an open, unrestrictive mind-set. This means allowing seemingly unproven concepts to live as possibilities, while acquiring additional (seemingly unproven) concepts and allowing them to also "live," until finally, the cumulative effect is undeniable.

This open-mindedness then results in a metaparadigm shift. This "larger view of reality" is the opposite of the simplistic reductionist thinking (taught in science classes) that attempts to isolate things to analyze them one at a time.

Personally, I see *Materialism* as a powerful scientific tool that gave us the great advances in our industrial economies during the nineteenth century. From my own experience as a naval architect and marine engineer, I know *Materialism* and classical physics are actually excellent *approximations* of reality. I personally used them to great advantage during the forty years of my working career. And while it is a wonderful *approximation* of reality, I also know that the *Materialist* model does not actually reflect the truth of our universe. Yet because *Materialism* is so widely endorsed in our Western culture, we need to understand both what *Materialism* is, and what it is not, in order to discuss consciousness.

1. MATERIALISM:
A. WHAT MATERIALISM IS:

Honest scientists realize that *Materialism* is only a perspective and do not hold it as a religious belief system (BS). They understand that *Materialism* is a perceptive device of assuming that matter (material) is all there is. It is only the pseudoscientists who misrepresent *Materialism* so religiously.

Historically, this seemingly cogent argument that matter is all there is goes back to the ancients, where some 2,350 years ago, the concept that the soul (or consciousness) was made of matter first appeared. As early as the fourth century BCE, Greek sage Epicurus (342–270 BCE) defined the soul as **"a body of fine particles."** He stressed that when the body loses breath and heat, the soul is dispersed and extinguished.

Later, the Roman poet Lucretius (99–55 BCE) followed on Epicurus, stating that *"the mind is composed of extremely fine particles."* Lucretius said that the relation between mind and body is so close that mind depends upon the body and therefore cannot exist without it. He made various arguments in support of his views, but also invented the idea that *if the soul were to be immortal, then why does it have no memories of its previous existence?* Lucretius' arguments, with the additional caveat that the mind is a function of the brain, were taken up and expanded in the nineteenth century by Thomas Huxley and others. So the *Materialist* belief isn't just 350 years old; it is truly archaic, being nearly 2,400 years old.

Unfortunately, the *Materialist* perspective also dictates the nonsensical concept that **consciousness somehow mysteriously arose from unconscious material...** In other words, the stones of the universe, given billions of years, eventually evolved into conscious beings... But just how does that work, exactly? Frankly, it sounds more miraculous than Jesus' miracles. Yet this "creation miracle" is a dogma religiously believed by all *Materialists,* especially by pseudoscientific *Materialist* skeptics.

B. THE FOUNDATIONS OF MATERIALISM:

Materialism is the seemingly self-evident concept that there is an objective reality "out there" that exists separate from our subjective observance of it. And that this reality is made out of substance which we call "matter." Our current widely accepted worldview of an objective reality made of matter is based on the science of ***rational empiricism*** and the principles of ***classical physics.*** which laid the groundwork for the *Materialist* paradigm.

RATIONAL EMPIRICISM: This is the belief that through experiments a scientist can learn about nature, and then using mathematics the scientist can describe and predict nature. This use of scientific reasoning to figure things out is called rational empiricism, which is one of the best ideas of all time.

REDUCTIONISM: This is the process of taking a thing apart to analyze each of its components, and it is the way that classical physicists approach their understanding of the world. Unfortunately, although many processes appear to be explicable by taking them apart, reductionism is also a limited idea, because within the boundaries imposed by classical physics (3-D plus time), it has difficulty discovering the emergent properties of systems. For example, if you have never seen an airplane and you find an old one and then you take it apart to determine what it was used for, you may never discover that it once flew. This same system tries to understand the world by taking it apart.

CLASSICAL PHYSICS: This is the logical paradigm that resulted from the data produced by the many experiments performed using rational empiricism during the seventeenth and eighteenth centuries, which culminated in the nineteenth-century perspective that supports the current cultural worldview of an objective material reality.

This materialist perspective rests on a foundation of six **First Principles**, which were presumed in the seventeenth and eighteenth century to simply be "self-evident." Yet to date, none have ever been proven to be true, and truth cannot be based on mere assumptions.

These unproven but assumed-to-be-true First Principle concepts are *reality, locality, causality, continuity, determinism, and certainty* as defined below:

> *REALITY:* This is the unproven presumption that asserts the physical world to be objectively real. This means that the observed world exists independent of our creative thought. In other words, this is a belief that we don't just conjure up the physical world in our thinking, but instead the physical world actually exists whether we are there to observe it or not. This concept is based on the next four concepts (locality, causality, continuity, and determinism) all being true, which they are now known not to be.

> *LOCALITY:* This is the unproven presumption that asserts that things can actually be separated by space and dimension. This carries with it the idea that objects can only be influenced by direct contact. Unmediated "action at a distance" is simply prohibited. Remember, this was a mechanistic concept invented before the discovery of radio waves and other forms of higher-frequency radiation, which all seem to operate *without* benefit of physical contact.

Also, we now have John Bell's *theory of non-locality,* which has been proven nearly a dozen times between 1972 and 2009 and entirely refutes any concept of locality. John Bell's theory is possibly the most important discovery of the twentieth century, more important than Einstein's theory of relativity. It proves conclusively that space and distance are truly illusions. And as a consequence, **it also proves that the concept of "locality," on which materialism is based, is entirely false.**

CAUSALITY: This unproven presumption asserts that the arrow of time points only in one direction and cannot be reversed; and assumes that cause-and-effect sequences are absolutely fixed. However logical this may seem to our perception, QED has shown that time can be measured in both directions, and further, that certain actions can be measured which show retro-causality. That is, observations made later can change what has been held in a quantum state from the past. Here we are back to the thought experiment of Schrödinger's cat, and in some instances, decisions made later can act with retro-causality. Several scientists have recently been able to show retro-causality in scientific experiments. So, this also disproves the idea that time only moves in one direction.

CONTINUITY: This unproven presumption asserts that there are no discontinuous jumps in nature, and that the fabric of space and time is "smooth." This means that there are no holes in the fabric of space and time. But at the micro-scale QED, nature is neither smooth nor continuous, and mostly exists only in a quantum state until it is observed by a living consciousness. In other words, it is one big vacuum until we observe it.

DETERMINISM: This unproven presumption asserts that things progress in an orderly, predictable way. Taken to its extremes this belief says that if we were able to know all the starting conditions, and to anticipate all the causal linkages that would unfold, then we could, in principle, predict the future completely. Unfortunately, this belief is based, too, on the validity of causality, reality, and certainty.

CERTAINTY: This is the unproven presumption that something can be proven to a mathematical certainty. In other words, "knowing for sure." However, Gödel's incompleteness theorems proved in **1931** that essential aspects of *certainty* could not be attained, and he found that:

> *"in any sufficiently powerful mathematical system, such as arithmetic, a statement that can be shown to be true, but that does not follow from the rule of the system will always exist."*

Thus, it became clear that the notion of a mathematical certainty cannot be reduced to a purely formal system. Gödel showed that such a system was not powerful enough for proving its own consistency, let alone that a simpler system could do the job. Also, Heisenberg had already restated this in **1927** as the "uncertainty principle."

Although it is comfortable to believe in these basic *First Principle* presumptions as being inviolable, all of them lack any basis in modern science because each of them, as explained below, has now been proven to be false by quantum electrodynamics and non-locality. They are still wonderful approximations, and very useful in the practical, applied science of engineering, but they are just not truths or facts.

This falsification, by QED, of the entire foundation of classical *Materialism*, while the same data was proving psi, is of extreme concern to *Materialist* believers. Proving the existence of psi would demand that they change all their comfortable belief systems (BS). So they are "deathly" afraid to look at any data that proposes to confirm psi (please excuse the pun). Another prominent *Materialist* believer, Douglas Hofstadter of Indiana University, said,

"If anything of (these) claims were true, then all the bases underlying science would be toppled, and we would have to rethink everything about the nature of the universe."

Hello? ... Better get busy rethinking then.

C. MATERIALISM MAKES SENSE:

"There is no question that adopting a materialistic perspective helped science historically break away from the constraints and biases (including censorship) of various religious institutions. Moreover, scientific methods based on Materialistic philosophy have been highly successful in not only increasing our understanding of nature and the universe, but, also in obtaining greater control and freedom through advances in technology. It is understandable how materialism became the cardinal assumption in mainstream science...

"Not surprisingly, when the assumption of materialism is questioned today, it typically evokes confusion and criticism, if not consternation, by conventional scientists... (But) a materialistic interpretation of the mind-brain relationship precludes the possibility that a greater spiritual/non-material reality could, in principle, exist... (Yet) It turns out that research on the cutting edge of consciousness science points strongly to the possibility that some sort of large Spiritual reality exists..."[15] Dr. Gary Schwartz, Ph.D., Chapter 37, *Oxford Handbook of Psychology and Spirituality*

These **First Principles** make sense to most of us. The chair I am sitting on looks solid enough, even though I know it is only an electromagnetic field that resists the electromagnetic field of my backside…and using this perspective, and the handy tool of classic Newtonian physics, has given us our modern industrial civilization. Consequently, it is automatically perpetuated by our system of higher education, where student scientists are automatically taught to analyze everything from the perspective of whether or not it agrees with these **First Principles** of Newtonian *Materialism*. Indeed, I spent my career designing ships using these principles of hydrodynamics and strengths of materials, all based entirely on *Materialism* and classical Newtonian physics.

On the other hand, having had a near-death experience forty-six years ago, while still an undergraduate college student, I know that the chair is just an elaborate illusion. I have also spent a lifetime observing how difficult it is for scientists trained in classical physics to set aside the common sense of *Materialism* in order to look "outside the box" of classical Newtonian physics to understand the further implications of quantum electrodynamics (QED)… But no matter how difficult it is to fathom, it is true that QED and non-locality have definitely shown consciousness is the underlying matrix of the entire physical reality.

For example, Dr. Albert Einstein, one of the greatest classical physicists and an honest believer in the common sense of *Materialism*, had great difficulty with the concepts of extrasensory perception (ESP) and was perplexed by the data produced by J.B. Rhine and others. Such data seemed to prove ESP was real, and being an open-minded scientist, Einstein stated:

> *"We have no right, from a physical standpoint, to deny* **a priori** *(out of hand) the possibility of telepathy."*

Relying on his classical training, he was unable to accept the data proving ESP and so continued to judge this data using First Principles of classical physics, and consequently he found it to be

> *"…suspicious that 'clairvoyance' yields the same probabilities as 'telepathy,' where the distance of the subject…from the sender has no influence on the result… This is a priori improbable to the highest degree, consequently the result is doubtful."*

In Einstein's experience and in accord with First Principles, the effectiveness of any force should decline with distance. Einstein saw this lack of diminishment as a clear indication that there must be something wrong with the data. But he was cautious enough to understand that history would record his words and thoughts, and so as an honest scientist, he did not deny the data.

Actually, rather than something wrong with the data, this was instead a clear indication that there is something (just as Dr. Alcock and Dr. Hofstadter both feared) *"horribly wrong"* with the foundations of the *Materialist* perspective.

Unfortunately, Einstein died in 1955, before John Bell's theorem of non-locality (1964) would establish, through numerous replicated experiments between 1972 and 2009, that space and distance are merely perceptive illusions.

Einstein's own theory of relativity had already shown, in 1915, that time itself was an illusion, but he was not yet ready to fully comprehend that space was also an illusion. Indeed, Einstein was profoundly uncomfortable with many of the claims of QED, which today are all commonly accepted by most physicists.

Acknowledging how Einstein quite reasonably suspected that the data was in error, and considering his desire for order, allows us to have sympathy for *Materialist* scientists schooled in this same perspective. Not unlike Einstein, today's scientists also have difficulty buying into the scientific data produced by paranormal experiments, but unlike Einstein, all too often, they naively refuse to look at the data proving the paranormal. Einstein, a true scientist, accepted the data and struggled to work with it.

Like Einstein, *Materialist* skeptics need to avoid future hubris by stopping there, at the point of saying *"We just don't know,"* rather than vehemently denying it. And further to realize that they may be the ones with the perceptive difficulty. Science itself, to remain alive and true science, MUST INCORPORATE ALL THE DATA. Intellectual honesty demands that all scientific data be placed into the paradigm, even if that requires completely rebuilding the paradigm from the ground up. It is only when small-minded *Materialists* dogmatically begin to call the data and the parapsychological scientists frauds and liars, merely because those they attack are able to see well beyond the limited vision of classical physics, that a problem sets in. Carl Jung described this *Materialist* mental deficit well:

> *"I shall not commit the fashionable stupidity of regarding everything I cannot explain as a fraud."* —Dr. Carl Jung 1919, in speech given to Society for Psychical Research (SPR), Kensington

MAYBE IT IS TIME TO PUT CONSCIOUSNESS BACK IN THE BOX: Since, according to Max Planck, this consciousness, which has been excluded from the limited box of physics for 350 years, is actually *"the matrix of all matter,"* maybe it is time for responsible scientists to figure out how to include consciousness *within* our modern models of the universe, even if that means expanding the box a great deal and refiguring everything.

D. HOW QUANTUM ELECTRODYNAMICS DISPROVES MATERIALISM:
Today, in 2016, medical science has extensive clinical data clearly proving, beyond all reasonable doubt, that consciousness outside the body exists, and that the afterlife also exists and is not a *religious myth*. Also, that qualification, "beyond all reasonable doubt," is the standard of proof required in all of science. So, from that aspect, the existence of the afterlife is now an **established scientific fact,** beyond all reasonable doubt.

Now, let's look at how QED has shown the foundational First Principles of materialism to be merely unproven archaic presumptions and are therefore no longer considered by any intelligent scientist to be scientific facts, but are today seen, by Dr. Carl Sagan's definition, as being merely religious superstitions.

These First Principles, repeatedly proven false in numerous scientific experiments, can no longer be assumed true. Here is how all these former conceptual "truths" (*reality, locality, causality, continuity,* and *determinism*) have been debunked:

- *Causality* was debunked by Einstein's ***theory of relativity***, which showed that the fixed arrow time was only an illusion, a mere misunderstanding caused by our assumption that time and space were absolutes. We now know empirically that WHEN something seems to happen is entirely dependent on the perspective of the observer. *There is no "when," only "now," and time itself is an illusion.* So, consequently, there can be no arrow of direct causality.

- *Continuity* was destroyed by the discovery that the fabric of quantum reality is, in fact, discontinuous, and only exists as an infinity of potentials at the small scales of subatomic particles, where space and time are neither smooth nor continuous, and everything remains in a quantum state until it is observed (it literally appears from the void when we need it). And also it may cease to exist as matter, when we (consciousness) stop observing it. However, this last premise is hard to verify simply because to verify it we need to observe it, which of course will precipitate it as matter again. But suffice it to say that reality is truly discontinuous.

- *Certainty.* Heisenberg's *uncertainty principle* took this concept out entirely when he presented his discovery and its consequences in a fourteen-page letter to Wolfgang Pauli in February 1927. **So, any form of *certainty* is an illusion.** And then again in 1931, Godel's incompleteness theorem showed that even a "mathematical certainty" couldn't be certain.

- *Locality* was also proven to be a perceptive error entirely by John Bell's theorem of non-locality (1964), which has now been proven by separate laboratory experiments nearly a dozen times between 1972 and 2009. *So, locality (space and dimension) is an illusion. We now call this new and opposite concept "non-locality."*

- *Reality* was based on the first four assumptions (*causality, continuity, certainty,* and *locality*) being true. So, because it is now proven conclusively that none of them are true, then the idea of an objective reality "out there" is also <u>entirely an illusion</u> since it no longer has any logical foundation. *There simply can be no "there" there.*

- *Determinism* has also fallen because it relies for its validity on the truth of the assumptions of *causality, reality,* and *certainty,* none of which exist in absolute terms any longer. So this presumption is also falsified.

So, with not one of the foundational presumptions or First Principles upholding *Materialism* being valid, then no thinking person can hold *Materialism* to be true any longer. This entire BS (belief system) or religion of *Materialism* has been completely falsified by modern science. Consequently, *Materialism,* which is based on those First Principles all being true, is an entirely false assumption completely debunked by QED and non-locality.

Any person, scientist or otherwise, who continues to believe in the dogma of *Materialism* is stupidly clinging to a comfortable BS that can now only be seen as a superstition, of as being their *religion.*

E. REDUCTIONISM IS ALSO DISPROVED BY QED:
The dissolution of all these assumed truths also challenges *reductionism,* which is the way that classical physicists approach their understanding of the world. Although many processes appear to be explicable by taking them apart, this mechanistic approach assumes that objects are separate (which *non-locality* proves is not true), and it also assumes that they interact in *deterministic* and *causal* ways (two concepts that have also been shown to be false). But if *locality, causality,* and *determinism* are false (as QED proves them to be), then this perspective that reductionism is a valid method of inquiry must also be false.

Consequently, any attempt to study something using *Materialism* and reductionism will, of course, give a completely false impression.

For example, take apart a Beethoven symphony and try to examine each note separately. You will never get the same impressions that Beethoven intended. Indeed, you won't even get close to the symphony when trying to comprehend one note at a time. Beethoven's Fifth must simply be taken as a whole to allow the epiphenomena to appear...da da da, dum.

WHY REDUCTIONISM DOES NOT WORK: Reductionism is a powerful tool in mechanics. But it is completely incapable of realizing the emergent properties (epiphenomena) of a system; to disassemble something is to disable it, if not kill it, and this is where reductionism fails entirely with dynamic systems.

> For instance, take water. Water is a molecule made of the combination of two gases, hydrogen and oxygen, which come together in the formula H_2O.

> If a chemist used the reductionist paradigm to analyze this combination of hydrogen and oxygen, he would first isolate hydrogen so that he could carefully study it. Then he would isolate oxygen and carefully study that. He may even accidentally discover that both gases are highly flammable when they accidentally come together and so burn down his laboratory.

> He will have to call in the fire department, which will use water (H_2O) to put out the fire that is oxidizing his hydrogen. But now for safety reasons he will keep the hydrogen and oxygen totally separate. So then he will never discover their emergent property, water, and he can never hope to find the second-level emergent properties: ice and steam. Consequently, he'll never get to the breakthrough he is seeking.

> No matter how extensive his notes on each individual gas, using reductionism he will simply never find the emergent properties of what the two gasses become when properly combined. Water is not at all like hydrogen or oxygen. Emergent properties are those that arise only when the two components are properly combined. He will continue to need this water to put out the hydrogen-generated fires, but as long as he isolates the gasses and studies them separately, his results will be completely unsatisfactory.

> It is only when the chemist thinks "outside the box" of reductionism and studies the two components together as a system called **water**

that he can find all the wonderful things water does: like form ice and vapor, rain, fog, et cetera. Inside the reductionist box, neither oxygen nor hydrogen is visible to normal human vision. But when they come together outside the perceptive box of reductionism so that the new emergent property of water arises, this emergent molecule becomes visible!

Chemistry long ago decided to use a systems paradigm rather than a reductionist paradigm. Chemistry wants to know what happens when we combine various chemicals into systems like H_2O, TNT, etc. Chemists are continually combining substances to find out what their emergent properties are. In that sense, chemistry is not reductionist science. Chemistry is instead all about the emergent properties. Chemistry is about combinations (systems) instead of components (reductions).

Reductionist thinking is about individual parts, which is why reductionist thinking will never be able to achieve a "Theory of Everything." To understand the whole enchilada, you have to think outside the reductionist box. Unfortunately, many reductionist-*Materialists* already know what they *want* to believe. Looking forward to retirement, they don't want to be bothered with disturbing little anomalies like out-of-body consciousness, or an afterlife, or spirit communication. Instead, their chosen *religion* is that consciousness could not possibly exist outside the body.

Here is the point where many *Materialists* often fall back on a statement by the late Dr. Carl Sagan: ***"Extraordinary facts require extraordinary proofs."***

Quoting Dr. Sagan, they point at parapsychological evidence proving the paranormal and then say that the normal level of scientific proof, **beyond a reasonable doubt,** is insufficient to prove the so-called paranormal. This is merely because the paranormal does not agree with their chosen *Materialist* worldview, and therefore (according to their BS) it should require extraordinary proofs. But this muddleheaded thinking entirely overlooks their own philosophical errors.

1. First, science is not a set of laws like the Ten Commandments set down in some holy book.
2. Second, their chosen perspective of archaic nineteenth-century materialism is not supported by the modern science of QED.
3. Third, they have no data to support their statement that consciousness survival is not true.
4. Finally, if extraordinary facts do require extraordinary proofs, then it is their own **extraordinary facts of *Materialism* that so completely disagree with modern science of QED, which would therefore require these extraordinary proofs.**

The so-called paranormal (psi) agrees with the modern science of QED and non-locality and should therefore be called "normal." Consequently, psi only requires normal levels of scientific proof. Instead, **Materialism itself should now be called "paranormal" or "abnormal,"** and *Materialism* should be now required to provide the "extraordinary proofs" because it has been scientifically proven that it is not "normal."

Of course, thinking scientists already know that such proof is most likely impossible because proving the underlying First Principles has already been shown to be impossible.

Interestingly, all the recent scientific evidence on the subject of an afterlife, coming from several separate streams of scientific inquiry, are yielding a single, very consistent perception. And that consensus is that **non-religious consciousness survival after death is a now a *proven scientific fact.***

2. QUANTUM ELECTRODYNAMICS:

A. WHAT IS QUANTUM ELECTRODYNAMICS (QED)?
The establishment of quantum electrodynamics changed everything. The same scientific breakthrough QED that allowed this shift away from the gross forces of mechanical propulsion and toward the subtle energies of electronics has, in just one hundred years, become the most proven theory in the history of science.

However, QED is also the least understood theory of all time. The reason for this lack of comprehension is a direct result of militant resistance from classical physicists who don't want to abandon their convenient and comfortable *Materialist* paradigm (BS) and so choose instead to purposefully ignore anything that does not agree with their First Principles.

QED says that time and space are illusions, and this means that the First Principles are also illusions. Water really does not seek its own level; it only appears to do that within the perspective-box of our three dimensions plus time. Non-locality shows that space and dimension are illusions, so there is no "level" for the water to be seeking. QED shows that physical reality is entirely an illusion, *"albeit,"* as Einstein said, *"it is a very persistent one."*

Strangely, those same *Materialist* scientists who do not want to accept the implications of QED are still quite happy using the electronics of their iPhones and the Internet, and relaxing with their plasma flat-screen televisions (all of which is impossible if the QED principles are not true).

Incoherently following some kind of cognitive dissonance, such *Materialist* "believers" will even use their cell phones (the operation of which proves QED and quantum strangeness to be true) to discuss with each other how both QED and quantum strangeness cannot be true, and how they are next going to try to prop up materialism and its underlying presumptions which form the bedrock of their classical physics worldview.

A complete explanation of quantum electrodynamics is provided in Chapter 11.

B. WHAT QUANTUM ELECTRODYNAMICS HAS ACCOMPLISHED:

As I have already stated, quantum electrodynamics has given us all of the following, which are inseparable from our modern life.
- Nuclear power, the atomic bomb, the hydrogen bomb, the neutron bomb
- Transistors, including all digital circuits on silicon wafer chips and diodes
- Computers all rely on that transistor technology
- Lasers, Blu-ray players.
- Functional Magnetic Resonance Imaging (fMRI)
- Digital photography
- Plasma flat-screen high-definition television
- Cell phones, CD players

Consequently, QED, as well as all its implications, must in fact be true.

C. WHY QED REMAINS MYSTERIOUS:

The mysterious part of QED is contained in the concepts of *superposition, complementarity, uncertainty,* the *"measurement problem,"* and *entanglement.*

Actually, each of these new terms of QED is just a different way of describing the single mystery: that a quantum object has no "definite" properties. It has no location in space or time. The only thing that it actually *has* is potential. QED has shown us that **the fundamental properties of the universe do not exist before they are actually observed by a living consciousness.**

The implications of that single thought are indeed the most awesome of all physics.

Therefore, our normal assumption that ordinary objects are entirely and absolutely separate is completely incorrect. And further, QED shows us that unmediated action at a distance is a *required* function which that consciousness uses to form our observed physical reality. Literally, the physical reality pops into existence as we arrive to look at it... And it (physical reality) is a constantly reappearing apparition, annihilating and remanifesting from the void (zero-point field) at the alarming frequency of twenty-three septillion times a second (every ten yoctoseconds).

But explaining that last paragraph would first require completion of some graduate courses in the physics of the zero-point field (ZPF) and the dark energy / light energy interface, and that is what we are trying to avoid here. Suffice it to say QED is real and it's why your cell phone works. And QED being real is what allows out-of-body consciousness and afterlife survival.

D. HOW QUANTUM ELECTRODYNAMICS ALLOWS CONSCIOUSNESS SURVIVAL:

Quantum electrodynamics requires an observation by a living consciousness in order to precipitate matter through the collapse of the wave function. In other words, *QED requires a consciousness outside of the physical reality to precipitate this physical reality*. Therefore, QED *requires* the existence and actions of out-of-body (outside the physical) consciousness, or there can be no physical reality.

Consequently, denial of out-of-body consciousness is denial of QED. So, if there is no consciousness existing out-of-body, then give us back those cell phones, and there will be no lasers, no transistors, no MRIs, et cetera.

On the other hand, if consciousness can exist out of the body, indeed is *required* to exist outside the body by QED, then your consciousness continuing after your physical body dies is not unusual; it is just another form of consciousness being out-of-body. Indeed, the afterlife is suddenly quite "normal."

This existence of the consciousness outside the body is further explained in Chapter 5, in the discussion of hypothermic cardiac arrest (HCA), where, despite all *Materialist* beliefs to the contrary, the latest clinical data shows that the consciousness of people who have been completely dead for several hours (and I mean literally and completely dead, with absolutely no respiration, no heartbeat, and no brain wave or EEG, just a cadaver in a cooler) are regularly and repeatedly brought back to life after undergoing HCA These highly specialized surgeries, which have been happening regularly now for over thirty years, result in "revived" patients who continue to live normal lives, with no loss or disability.

But more interestingly, many of them (around 20 percent) also have near-death experiences during the three or four hours while they were clinically "completely dead." Now, obviously, if they have recorded memories while they were completely dead (by all medical definitions of what constitutes "dead"), then *their consciousness was surviving somewhere outside of their "dead"*

50

body and outside their non-operating brain, for those several hours of not breathing and not having a heartbeat and no brain function.

This is all entirely factual, but it is so unbelievable to a *Materialist* believer who already "knows" this is all "impossible" that I can imagine some of them putting the book down and looking around for a stiff drink. Instead, if they actually want to claim to be scientific, they should check out all the data I am freely pointing out (like the eighty-seven videos on YouTube about HCA) and stop foolishly and fearfully hiding from the facts. It is simply time to grow up and face the music.

Again... Hello?

Regardless, it is that "somewhere" (where these consciousnesses belonging to the completely dead cadavers hang out during the interim three or four hours) which is, by simple definition, an "afterlife."

So, this thirty years of medical practice with HCA proves that consciousness can and does survive in an afterlife. Many of these HCA patients recall exactly where they were and what was happening during the time (three to four hours) when their brains were completely dead, without blood or oxygen.

Further, this process of bringing people back from the dead is entirely replicable in any hospital with the proper medical procedures available. And it has been frequently replicated, as I fully discuss later in Chapter 5. Further, it is beginning to be available in nearly every major city. We now have it (since late 2014) in my hometown available at San Francisco General Hospital.

The smaller outlying hospitals don't have HCA. But if you will watch the videos at these URLs and read Dr. Sam Parnia's book, you will see that patients dead and not breathing for forty minutes and longer have been brought in, cooled down, and then using HCA techniques brought back to life with no loss of function. So maybe the outlying hospitals don't need it, so long as it is available and they can get the cadaver over to the general hospital within a few hours of death. In any case HCA has become so popular that YouTube had to create a special playlist for the most popular eighty-seven HCA videos. So enjoy these mostly two- to four-minute shorts, but fully verify all this by checking out the data (what a novel idea for a *Materialist* believer...actually looking at the conflicting data).

- **Popular Videos - Hypothermia & Cardiac Arrest - YouTube**
  87 **videos**
  https://www.youtube.com/playlist?list=PLzkneUrc2KJjlhlr-
  TOHNdScanURwebI1U

[lb] *Erasing Death,* Dr. Sam Parnia's book, also explains this in great detail.

3. BIOCENTRISM

A. WHAT IS BIOCENTRISM? HOW WAS IT ESTABLISHED?

Biocentrism is a recently evolved worldview being championed by frontier biologists, which fully explains how this *"mind as the matrix of all matter"* that physicist Max Planck spoke of can be understood. And more importantly, biocentrism is completely compatible with consciousness survival outside the body during out-of-body experiences (OBEs), like the near-death experience (NDE). It is also completely compatible with consciousness survival in an afterlife, subsequent to the death of the physical body.

Biocentrism is the perspective that life (consciousness) is fundamental to the universe, and precedes everything else. But it is by no means a new idea. It is the same idea that the Nobel laureate physicists who brought us QED believed in. The conclusions of biocentrism are based on mainstream science, and it is a logical extension of some of our greatest scientific minds.

This concept of biocentrism was alluded to by Henry P. Stapp, of the Theoretical Physics Group at Lawrence Berkeley National Laboratory, University of California, Berkeley, in a paper that explains how John von Neumann's "Process 1" allows for the compatibility of QED with consciousness survival (it may be a bit technical at first but please bear with me).

> *"Orthodox quantum mechanics (QED) is technically built around an element that (John) von Neumann called Process 1.*
>
> *"In its basic form, Process 1 consists of an action that reduces the prior state of a physical system to a sum of two parts, which can be regarded as the parts corresponding to the answers 'Yes' and 'No' to a specific question that this action poses, or 'puts to nature.' Nature (consciousness) returns one answer or the other, in accordance with statistical weightings specified by the theory.*
>
> *"Thus, the standard statistical element in quantum theory enters only after the Process 1 choice is made, while the known deterministic element in quantum theory governs the dynamics that prevails between the reduction events, but not the process that determines which of the continuum of allowed Process 1 probing actions will actually occur. The rules governing that selection process are not fixed by the theory in its present form.*
>
> *"This freedom can be used to resolve in a natural way an apparent problem of the orthodox theory its BIOCENTRISM. <u>That resolution produces a rationally coherent realization of the theory that preserves the basic</u>*

orthodox structure, but allows naturally for the possibility that human personality may survive bodily death."
—Henry P. Stapp, *Compatibility of Contemporary Physical Theory with Personality Survival*

Now, that last quotation is a little deep, but if you work with it long enough, you will see that what Henry Stapp is saying can be paraphrased as follows:
In quantum electrodynamics, nothing exists until the observer makes a choice, and biocentrism, through its reliance upon a living consciousness to make this choice, allows the distinct possibility that consciousness exists entirely outside the physical and therefore QED (as interpreted by biocentrism) allows for consciousness survival in an afterlife. —The author, Dr. Alan Hugenot

B. WHY BIOCENTRISM WORKS:
Biocentrism allows for consciousness, or mind, to be Max Planck's *"matrix of all matter."* All that is required to understand this hypothesis is to shift your understanding to consciousness being the smallest particle, instead of it being a particle made of substance. This agrees with the QED statement that matter is not made of matter, but is instead a field. Simply consider that electromagnetic field as being a conscious thought.

C. WHAT BIOCENTRISM HAS ACCOMPLISHED:
This simple change of perspective (biocentrism) has allowed major shifts in the thinking of biologists and cell biologists, for example:

MICROBIOLOGY & CONSCIOUSNESS: According to the latest theories in the cell biology of epigenetics, DNA does not contain hereditary material itself, but is more like a barcode. DNA is capable of receiving hereditary morphogenic (formative) and person-specific information from the environment (i.e., non-local consciousness). The reciprocal information transfer takes place between the morphogenic field and the living cell structures via resonance with specific frequencies, even at the smallest sub-cellular level of electron spin resonance and nuclear magnetic resonance (quantum spin correlations).

This creates the postulate that:
Physical reality is continually supported and recreated by the underlying non-local consciousness which exists at the sub-nuclear level of the body's cells. But is not contained within the cells themselves.

Erin Schrödinger postulated in 1944 that the DNA of living organisms has receptor or resonance potential for receiving and recording information from non-local space. Also, because there is not enough room in the cells to store this vast amount of information, it must come from outside the cell, and come instantaneously. So that can't happen through normal protein metabolism because it is too slow.

DNA is the only "part" of the cell that remains constant, while each cell is constantly being replaced and must instantaneously reconstruct itself to its original blueprint (twenty-three septillion times a second). And because each entire cell is destroyed, the blueprint cannot be within the cell; instead the cell must be in continual contact with the surrounding morphogenic consciousness (its immediate environment) and the information routed by the DNA in the cell nucleus. The unique environment appears to hold the specific information or morphogenic consciousness which that unique cell needs to develop and specialize. Dr. Bruce Lipton, Ph.D. covers this topic quite well in his several recent books.[16,17]

BEYOND TIME AND SPACE: Dr. Robert Lanza is an expert in regenerative medicine and scientific director of Advanced Cell Technology Company. Renowned for his extensive research that dealt with stem cells, he was also famous for several successful experiments on cloning endangered animal species. Dr. Lanza recently published a book, *Biocentrism: How Life and Consciousness Are the Keys to Understanding the Nature of the Universe*, wherein he states that life does not end when the body dies. He calls this new theory *biocentrism*.

BIOCENTRISM'S TAKE ON THE COSMOS: Dr. Lanza set down seven principles that define the new paradigm of biocentrism (The statements in italics are direct quotes from his above book):

> *FIRST PRINCIPLE: What we perceive as reality is a process that involves our consciousness. An external reality, if it existed, would (by definition) have to exist in space. But this is meaningless, because space and time are not absolute realities. They are instead tools of the human and animal mind.*

> *Non-locality* proves that space is only an illusion, so there is no space for the absolute reality to be located in.

> *SECOND PRINCIPLE: Our external and internal perceptions are inextricably intertwined. They are different sides of the same coin and cannot be divorced from one another.*

> *THIRD PRINCIPLE: The behavior of subatomic particles (indeed all particles and objects) are inexplicably linked to the presence of an observer. Without the presence of a conscious observer, they at best exist in an undetermined state of probability waves.*

This is simply basic quantum electrodynamics, regarding the observer principle and the state vector collapse, and is basic first-year science for all physicists.

> **FOURTH PRINCIPLE:** *Without consciousness, "matter" dwells in an undetermined state of probability. Any universe that could have preceded consciousness only existed in a probability state.*

> **FIFTH PRINCIPLE:** *The structure of the universe is explainable only through biocentrism. The universe is fine-tuned for life, which makes perfect sense as life creates the universe, not the other way around. The universe is simply the complete spatiotemporal logic of the self.*

In his book Dr. Lanza explains the more than two hundred physical constants that had to be exactly within certain very narrow limits for life to have evolved at all. The coordination of these two hundred parameters could not have been random. Two hundred factorial is incalculable by my computer. Ten factorial is 3.6 million to one, yet twenty factorial is 2.4 billion-billion to one or 2.4 quintillion to one, so you can see that the two hundred becomes an incalculably large number.

> **SIXTH PRINCIPLE:** *Time does not have a real existence outside of animal-sense perception. It is the process by which we perceive changes in the universe.*

> **SEVENTH PRINCIPLE:** *There is no absolute self-existing matrix in which physical events occur independent of life. This is because space like time is not an object or a thing. Space is another form of our animal understanding and does not have an independent reality. We carry space and time around us like turtles with shells. Thus, there is no absolute self-existing matrix in which physical events occur.*

Again this brings us back to Max Planck's truth that *"Mind (consciousness) is the matrix of all matter (reality)."*

D. HOW BIOCENTRISM ALLOWS FOR CONSCIOUSNESS SURVIVAL:
This biocentrism theory implies that death simply does not exist. It is an illusion that arises in the minds of people. It exists because people identify themselves with their physical bodies. They believe that the body is going to perish, sooner or later, and think their consciousness will also disappear along with the physical body.

However, consciousness, in fact, exists outside of constraints of time and space, and outside the physical body. It is able to be and act anywhere: in the human body and outside of it. That idea fits well with the basic postulates of quantum electrodynamics science.

Dr. Lanza also believes that multiple universes can exist simultaneously. This multiple universes theory is not supported by his data, although it is compatible with his data. These universes contain multiple ways for possible scenarios to occur. In one universe, the body can be dead. And in another it continues to exist, absorbing consciousness, which migrated into this universe where the body is still alive.

His extrapolation into multiple universes is for me, personally, a bit of a stretch, but I understand that it is just Dr. Lanza trying to explain all the data. Like Einstein trying to explain all the data to us using relativity. However, there is no scientific need (no scientific question) answered by this extension into multiverses. So, applying Ockham's razor, I don't go there.

Briefly, this multiple universe theory implies that a dead person, while traveling through the same tunnel, ends up not in hell or in heaven, but in a similar world he or she once inhabited, but this time alive. And so on, infinitely. But as I don't believe Lanza has experienced a personal NDE to confirm his speculation, there is no need to go further than we can know.

From my own empirical experience of an NDE and my work with after-death communications (ADC) as a medium, I can easily agree with the first half of Lanza's theory. But using the best science I cannot buy into his unsupported secondary explanation with the multiple universes. Now, neither can I say that it isn't so.

A complete explanation of biocentrism is given in Chapter 12.

But now let's look at some additional evidence for biocentrism that comes from beyond the grave, in communications received over a hundred years ago.

E. PRECOGNIZANT PREDICTIONS of BIOCENTRISM (1906):
Back in 1906, just before Einstein's theory was published, a New York lawyer, Edward C. Randall, published his findings after fifteen years of work with the direct voice medium Mrs. Emily French, where he had held after-death discussions with Dr. David C. Hossack M.D., founder of Columbia Medical School. Dr. Hossack, who died at age sixty-six on December 22, 1835, told Randall seventy-one years later in 1906:

> *The spiritual plane is filled with ether* (dark matter) *similar to earth substance* (light matter) *but in a very high state of vibration... The universe is all material substance or matter in different and varying states of vibration.*" —Edward C. Randall[18]

This is a statement of the alternative dimensions postulated by string theory and M-theory which would be hypothesized by physicists in the late twentieth century. Further, Dr. David. C. Hossack[19] told Randall that:

"The most learned scientist among the inhabitants of earth has practically no conception of the properties of matter, the substance of the universe, the visible and the invisible."

In other words, "Your scientists haven't yet discerned dark energy and dark matter, and so have practically no conception of the properties of the universe that actually matter." But Dr. Hossack went on to say:

"I did not (know this either) *when I lived among you though I made a special study of the subject. That, which you see and touch making up the physical or tangible, and having three dimensions (3D), is the lowest or crudest expression of life force and not withstanding my long study of the subject, the idea that the physical (matter) had a permanent life form or that what you call space was composed of matter filled with intelligent and comprehensive life* (conscious universe) *in a higher vibration* (separate energy field) *never occurred to me."* [20]

Here are some videos discussing Dr. Lanza's biocentrism theories:

- **Dr. Robert Lanza Discusses Biocentrism & Stem - YouTube** https://www.**youtube**.com/watch?v=4_LU0JG5I8M Jan 9, 2013—Uploaded. Biocentrism & Stem Cell Research

- **Dr. Robert Lanza on Theory of Biocentrism - YouTube** https://www.**youtube**.com/watch?v=zI_F4nOKDSM Jun 7, 2011—Uploaded. This is part one of Robert **Lanza's** talk on biocentrism at the Science and Nonduality Conference.

- **Dr. Robert Lanza at CCRI, 10/29/2013 - Part I - YouTube** https://www.**youtube**.com/watch?v=7z61aAmTpIk Dec 6, 2013—Uploaded. **Dr.** Robert **Lanza**, Chief Scientific Officer, Advanced Cell Technology.

- **Dr. Robert Lanza Part I (StemConn 2015) - YouTube** https://www.**youtube**.com/watch?v=W5I2xg_ct8Q Apr 28, 2015—Uploaded. Why is Robert **Lanza** not funded.

- **Biocentrism - Robert Lanza - YouTube** https://www.**youtube**. com/watch?v=YehIxgLNIJg Nov 12, 2014—Uploaded

- **Dr. Robert Lanza, 3/24/13 - YouTube** https://www.**youtube**. com/watch?v=sf7NCVqh0EE Mar 24, 2013—Uploaded.

CHAPTER 5

SCIENCE OF HYPOTHERMIC CARDIAC ARREST (HCA)

"There is no Death. There are no dead."
Modern Spiritualism

A. HOW LONG CAN A PATIENT BE DEAD AND STILL RETURN TO LIFE?

We have all been told by *Materialist* fatalists that *"when you're dead, you're dead."* But as is the case with most of their prattle, this is also just not true, and never has been.

Dr. Sam Parnia, M.D., Ph.D., who divides his time between hospitals in New York and London, describes in his book *Erasing Death* his work in ICUs resuscitating patients who have died through cardiac arrest. He often brings these patients back to life—patients who have been completely dead, with no respiration and no heartbeat, for three or four hours. This seeming "miracle" is made possible because of his equipment. He can provide medical procedures known as *Therapeutic Hypothermic Cardiac Arrest (HCA)* that are only now (in 2016) becoming available in most of the world's major hospitals.

With the proper equipment immediately available, Dr. Parnia or any physician trained in the technique can quickly lower the temperature of a patient's dead cadaver to hypothermic levels. This cooling action preserves the body's cells, preventing the decay that normally begins immediately at death as the normal body temperature of 98.6°F slowly declines.

When a patient experiences a cardiac arrest, Dr. Parnia can prevent decay of the body's cells simply by immediately lowering the body temperature to hypothermic levels, below 93°F (usually 89°F is the target temperature). When respiration and blood circulation of oxygen to the body's cells stops, if those cells are not immediately cooled, then the temperature of the now dead cadaver, which starts at 98.6°F, will slowly drop at the rate of about 2°F per hour. Left alone, it will take nearly three hours for the cells to cool down to safe non-deteriorating levels below 93°F, and in those three hours, sufficient deterioration will already have occurred to prevent the body from ever being revived.

On the other hand, when the equipment is available, the person having the cardiac arrest can be stabilized and placed "on ice," and left there, while the surgeons fix the problems with the heart.

Unfortunately, standard procedure at most hospitals without HCA available does not include this emergency cooling capability. So instead, they will immediately apply more primitive procedures of cardiopulmonary resuscitation (CPR) followed up with the use of defibrillation (electric shock paddles) to restart the patient's heart. But this obsolete procedure of CPR without repairing the heart first, and merely hoping that the broken heart can limp along until the necessary repairs can be made during a follow-up surgery (tomorrow or next week), has a low survival rate (below 15 percent). Yet, most hospitals do this because they want to keep the blood and oxygen flowing to those cells by whatever means possible. Unfortunately, these primitive procedures ignore the fact that the original problem, which caused the heart attack in the first place, has not yet been repaired.

Normally, the heart will have stopped because it had a blocked artery and was not receiving any blood. If the blockages in the blood flow to the heart muscle, which are causing the heart attack to begin with, are not resolved before the heart is restarted, then the heart muscle is being forced to continue pumping while it is still not getting any oxygenated blood, and that will literally kill the heart muscle.

Consequently, CPR and defibrillation are only very primitive attempts to force the unrepaired heart to go back to work while it is still having a heart attack! This is about as smart as forcing a patient to walk on a broken leg without crutches. Such extreme measures can only cause more destruction to the heart muscle, and almost always fail.

Notably, since the 1970s, Baylor Medical Center in Houston, Texas, has been developing the techniques of resuscitation using hypothermic cardiac arrest (HCA).

A QUARTER CENTURY AGO – STONE-COLD DEAD FOR AT LEAST TWENTY MINUTES: Only a few hospitals have the equipment to do this specialized surgery, so until now it has not received much media attention. However, a quarter of a century ago, back in August of 1991, at the Barrow Neurological Center in Phoenix, Arizona, one very completely and carefully documented case of HCA was reported, and because the patient had a near-death experience while she was (by all our markers) completely dead, both she and the doctor were later interviewed on CBS television's news show *48 Hours*.

A team of doctors led by Dr. Robert Spetzler, M.D. at the Barrow Neurological Center had placed Pam Reynolds[21] into full hypothermic cardiac arrest (HCA). They dropped her core temperature down to 60°F (16°C) and completely drained all the blood from her head and left her with no measurable breathing, heartbeat, or brain waves. She was completely "stone-cold dead" for a little under one hour. But during this time she was hooked up to every medical measuring device available, measuring blood pressure, blood flow, oxygen level in her blood, core body temperature, and cortical brain activity, all of which verified that she was actually dead by every definition we have for death. This story is told in all of Chapter 3 of Dr. Michael Sabom, M.D.'s book, *Light and Death* and also on p. 220 of Chris Carter's book, *Science of the Near-Death Experience.*

While Pam Reynolds was completely dead, clinically verifiable because of the instruments she was connected to, she began and progressed through a typical near-death experience, watching the surgeons from the top of the room, moving through a tunnel into the light, where she met with her deceased grandparents and a great-great aunt. Then when it was time to return, she went back through the tunnel and reunited with her cadaver. Her heart was stopped with massive doses of potassium chloride at 11:05 a.m. By 11:25 she had a body temperature of 60°F (16°C), and no EEG brainstem activity, despite the 110 decibel clicks in her ears. At this point the cardiopulmonary bypass machine was shut down and her blood was completely drained out. So, now she had no heartbeat, no circulation, no breathing, and no brainstem activity. She was literally dead from 11:31 a.m. until at least 11:52 a.m., a period of over twenty minutes. The aneurysm was removed and by noon they were circulating her warmed blood using the cardiopulmonary bypass machine. Her EEG was coming back and by 12:32 p.m. her temperature was again 89.6°F (32°C). The cardiopulmonary bypass machine was turned off, and she was officially alive again.

What makes the operation even more spectacular is that while Pam was dead, she had a near-death experience, which she also describes in both books referenced above. Her description includes her watching the surgeons during the time when she was "stone-cold dead." She is in the top of the room watching the surgeons perform this surgery.

B. HYPOTHERMIC CARDIAC ARREST - COMPLETELY DEAD for THREE to FOUR HOURS:

But twenty-four years later, Dr. Sam Parnia does this same procedure regularly in New York City and in London, UK, during cardiac arrests. He has also refined the technique so that after cooling the deceased cadaver to a temperature between 89.5°F (32°C) to 93°F (34°C), Dr. Parnia can leave the patient in "cold storage" (completely dead) for up to four hours, while attending physicians use fMRI and other techniques to find out what is causing the problem with this patient's heart and make the necessary repairs, and with plenty of time before they have to start the patient up again. This new therapy is now spreading rapidly to American hospitals, and those that have it are busy marketing it to the public as a miracle. Those without it are trying to keep that fact hush-hush.

The bare facts are that when doctors are given sufficient time like this to solve the problem before restarting the heart, they can avoid killing the heart's muscle tissue by not forcing it to work without fuel and overtaxing its already dwindling capabilities. Doctors can take the time to locate the clot using an fMRI scan, design a treatment to remove the clot, install a bypass or a shunt, and get the heart ready to function properly again with the capability of full blood flow before it is restarted using the defibrillation paddles.

NEW APPLICATIONS OF THIS TECHNIQUE: Since Dr. Parnia published his book *Erasing Death* in December 2013, American hospitals all over the country are suddenly rushing to institute this procedure before their local market discovers that they don't have this inexpensive and simple-to-apply therapy. Now, as it is becoming standard practice, new procedures are being added on top of Dr. Parnia's work. For example, when patients are brought in dead on arrival (DOA), it has been found possible to rapidly cool them down, even as much as forty minutes later, and then slowly bring them back. Within forty-eight hours they may be "back to normal" and ready for release from the hospital. This is illustrated in the following video.

- **Therapeutic Hypothermia for Heart Attack Victims.**
  Uploaded on Oct 23, 2009.
  Doctors at Penn Medicine's Center for Resuscitation Science have developed an innovative technique for bringing heart attack victims back to life. Listen to Dr. Lance Becker and others discuss the use of therapeutic hypothermia to help save lives after cardiac arrest. To learn more, please visit
  http://www.med.upenn.edu/resuscitation/   or
  http://www.pennmedicine.org
  https://youtu.be/vTHvUs_eA7g?list=PLzkneUrc2KJjlhlrTOHNd-ScanURwebI1U

Here is another video on this "miracle" process. Please note when the physician in that video says, *"The window of opportunity is three to six hours after cardiac arrest..."* Then think about what his words literally mean. *If you bring the cadaver in within three to six hours after the patient died from cardiac arrest, we can bring them back using hypothermia...* After that they are probably gone.

- **Therapeutic Hypothermia –Piedmont Healthcare:** Here is one of the best videos on this subject.
  https://youtu.be/ROZpccPrFzglist=PLzkneUrc2KJjlhlrTOHNd-ScanURwebI1U

WHY ISN'T THIS PROCEDURE AVAILABLE IN MORE HOSPITALS?

One is just driven to ask: If that first video is seven years old, and it says that this procedure was available in most European hospitals at that time (2009), why is it 2016 before it has started to be available in America?

Why haven't we heard about this new procedure, which has been used successfully for a quarter century?

And more importantly, why does American medicine drag its feet on life-saving new technologies?

Why isn't every hospital immediately acquiring this cooling equipment for their ICUs?

The short answer is that most medical doctors are schooled in the precepts of *Materialist* dogma. Most don't even yet believe in NDEs; they are comfortable with their *Materialist* perspective and reductionist medicine and so want to explain NDE as the delusions of a dying brain suffering oxygen starvation, or hallucinations from anesthesia. Instead of listening to those who have been there and done that.

Arrogantly, they still insist that their "higher learning" makes them experts even on things they don't know about. It is still difficult for them to accept this "out of the box" HCA therapy. But the therapy is so miraculous that they are being forced to allow it, or be left behind in the gold rush. As they implement it, to save face, they are also forced by its very acceptance as a therapy to validate that NDEs are real and out-of-body consciousness is real.

This cooling of the cells to prevent damage is the most revolutionary advancement in resuscitation medicine of the last thirty years. And by simply adding it to a hospital's available tools, many deaths from cardiac arrest can simply be reversed through cooling the body down to between 89.5°F (32°C) to 93°F (34°C). Although, exacting temperature control is vital, because it has also been found that when the body temperature dips below 89.5°F (32°C), other complications set in.

Since Dr. Parnia's book was published in December of 2013, HCA has become one of the hottest new therapies in America. Finally, this thirty years of medical practice with HCA, which proves that consciousness can and does survive in an afterlife, is being implemented nationwide. HCA has become so popular that YouTube had to create a special playlist for the most popular 87 HCA videos. So enjoy them they are mostly 2-4 minute shorts, but fully verify what I am saying.

- **Therapeutic Hypothermia for Heart Attack Victims**
  Uploaded on Oct 23, 2009
  Doctors at Penn Medicine's Center for Resuscitation Science have developed an innovative technique for bringing heart attack victims back to life. Listen to Dr. Lance Becker and others discuss the

use of therapeutic hypothermia to help save lives after cardiac arrest. https://youtu.be/vTHvUs_eA7g?list=PLzkneUrc2KJjlhlr-TOHNdScanURwebI1U

- **Therapeutic Hypothermia –Piedmont Healthcare:** Here is one of the best videos on this https://youtu.be/ROZpccPrFzglist=PLzkneUrc2KJjlhlrTOHNd-ScanURwebI1U
- **Popular Videos - Hypothermia & Cardiac arrest - YouTube 87 videos** https://www.youtube.com/playlist?list=PLzkneUrc2KJjlhlr-TOHNdScanURwebI1U
- *Erasing Death*, Dr. Sam Parnia's ©2013 book also explains this in great detail.

C. WHAT HAPPENS TO INDIVIDUAL CONSCIOUSNESS AT DEATH?

As all these proofs of consciousness survival begin to collate, the question arises: *Okay then, what actually happens when we die and go to the next dimension?*

The energy of our consciousness simply cannot cease to exist, even in strict accordance with the laws of classical Newtonian physics (*Materialism*). If the consciousness existed within the Newtonian paradigm of 3-D plus time (which it does not), then its energy could not disappear at death and it would have to go somewhere else within the box. The Law of Conservation of Energy states that *energy never disappears; it only changes form, with no beginning or ending.*

So, from that perspective, *Materialism* itself requires consciousness survival through energy conservation.

But consciousness actually exists outside the limited perspective box of 3-D plus time (*the explicit order*). It lives within the invisible *implicate order*. So there is nothing to measure because it does not exist within the physical realm that we can discern (or measure).

Dr. Pim van Lommel, M.D., a Dutch cardiac physician and personal acquaintance, has spent fifteen years studying near-death experiences and looking at cardiac arrest patients who had an NDE as well as those who did not. He examined the history of their vital signs in the ICU during the cardiac arrest and also interviewed them after the fact. Dr. van Lommel summarizes the current state of survival theory among scientists collating quantum mechanics and the NDE:

> *"Most people still believe that death is the end of everything.... That used to be my own belief. But after many years of critical research into the stories of NDErs, and after careful exploration of current knowledge about brain function, consciousness, and some basic principles of quantum*

physics, my views have undergone a complete transformation. I found the most significant finding to be the conclusion of one NDEr: 'Dead turned out to be not dead.' I now see the continuity of our consciousness after the death of our physical body as a very real possibility."

Here is a video URL:

- **Consciousness and Near-Death Experiences**—Dr. Pim van Lommel. Uploaded Oct 13, 2014

Pim van Lommel details his research into near-death experiences, the scientific and spiritual implications of consciousness independent of brain functioning. https://www.youtube.com/watch?v=RkF4KzWTrKA

Obviously, if the consciousness exists in that 96 percent of the universe that *Materialist* physics currently describes as **dark energy** and **dark matter,** which our best *Materialist* physics cannot even discern with any of our crude measuring devices, it also is obvious that consciousness does not have to follow any of these *Materialist* **First Principles,** which have been debunked by QED and non-locality.

CHAPTER 6

SCIENCE OF NEAR-DEATH EXPERIENCES

And My Own Story, Getting Involved in Studying Death & Beyond:
Excerpted from my earlier book, *The Death Experience—*
What it is like when you die

"If they say to you 'where did you come from?' say we came
from The Light." Jesus, quoted in the Gospel of Thomas

"In a moment, in the twinkling of an eye we shall be raised incorruptible
and we shall all be changed." St. Paul 1 Corinthians 15:51-52
Quoted from the lyrics of Brahms'
Requiem (1866/68)

The study of the near-death experience was formalized by the International
Association for Near Death Studies (IANDS) when it was formed in 1978 as
a response to the interest generated by Dr. Raymond Moody's 1975 book,
Life After Life. The association was formed by a number of concerned medical doctors and, in the years since, has grown to become a worldwide scientific body that also assists the NDE survivors with the psychological aspects of
coming to grips with what has happened to them.

NOTE: Being a professional member of the association since 1997,
as of 2015, I currently serve as a member of the IANDS International Board of Directors.

Since that time a great deal of anecdotal evidence has been collated and
compared, which includes thousands of NDE narratives. This empirical evidence all agrees with the medical evidence and data now available from
resuscitation technology. This agreement between the empirical (anecdotal)
evidence and the resuscitation data should now silence all the superficial
conjecture and unsupported speculations about NDEs being hallucinations
of an "oxygen-starved, dying brain."

Yes, from my own experience I know that it really does happen instantaneously, literally *in the twinkling of an eye.* Just as St. Paul knew from his own NDE (see 2 Corinthians 12 where Paul describes his own NDE).

The so-called death is merely a paradigm shift, a complete change of viewpoint. Those left here may call it death while looking at a vacated cadaver on this side, but no one ever actually dies. What occurs instead is that we merely change our perspective from this habitual frame of reference in 3-D light energy, to a perspective contained in alternative dimensions of dark energy located in non-local space, where the dense hulk of a cadaver is no longer needed, and so we simply leave it behind, allowing our spirit to expand. Personally, my having done that, dying, going out-of-body and then reincarnating back into the same body, changed everything for me. Here is my personal story of that experience.

1—MY PERSONAL NEAR-DEATH EXPERIENCE (1970 AD):

It was in May of 1970, and I was in Eugene, Oregon, attending college on the GI Bill after having served in the Navy during the Vietnam War. I was attending evening classes at the University of Oregon and working days as a draftsman/engineering technician, designing fire sprinkler systems for public buildings. I owned a motorcycle and commuted from the university campus out to my job in West Eugene.

Up until that time, I had always been a pretty normal American. My family attended Protestant services on Sundays, I liked hiking and camping with the Boy Scouts. I got good grades in history, math, and science. I joined the Navy, went to boot camp in San Diego, learned ship navigation at Quartermaster's school in Newport, Rhode Island, and enjoyed navigating my destroyer to foreign ports during the Vietnam War, before returning home to start college.

My life up to that point had been mostly concerned with finding a career, getting a date, being accepted at a college, paying for it, et cetera. Consequently, any questions about what happens after death and whether our consciousness survives were pushed aside. I easily accepted the prevailing Western cultural materialist philosophy that we live only one life, and if asked what I believed came after death, I would have said, "**According to my church we go to heaven or hell, but I'm not actually sure what really happens.**" At that stage of my life there were many more pressing problems to solve, and thoughts of the afterlife could wait.

Cyclist listed as 'critical' after accident

A motorcyclist was hospitalized in critical condition Monday afternoon after he struck the side of a moving car on West 11th Avenue near City View Street.

Injured was Alan Ross Hugenot, 21, of 2745 Emerald St, Eugene. A spokesman at Sacred Heart General Hospital in Eugene said Hugenot has head injuries and multiple fractures.

Hugenot remained in critical condition Tuesday morning, but the hospital said he had shown "some improvement."

The accident occurred about 4:50 p.m. as the cyclist was traveling east on 11th Avenue. Eugene police said a westbound car turned in front of Hugenot and Hugenot was unable to avoid colliding with the side of the car.

Officers cited Floyd Hargrove, 46, of 3356 Royal Ave., Eugene, on a charge of making an illegal left turn.

Article from May 29th 1970 *Eugene Register Guard*.

But everything changed on the evening of Monday, May 27, 1970. I was riding my motorcycle back to the University of Oregon campus on the way home from my design job, when I was involved in a collision with a car that turned left in front of me, and left me nowhere to go. I was severely injured and rushed by ambulance to the intensive care unit at Sacred Heart General Hospital, Eugene, Oregon, where I was hospitalized for about thirty-three days.

My right femur (thigh bone) was shattered, a portion of my right wrist (radius) was broken off and had lodged near my elbow, my right kneecap was completely crushed, and my head had experienced extreme trauma in a violent battering. My upper teeth had been destroyed entirely on one side of my jaw, and there were numerous fissures in my skull above the upper teeth, caused when my open mouth and skull collided with the luggage rack on the car.

When I arrived in the ER, my condition was *critical*, but I had been stabilized and was not hemorrhaging. So, they began preparing me for surgery. However, just prior to surgery, and about three hours after the accident, I became incoherent and unconscious.

My head injury had caused my brain to begin swelling, and in those days, they did not have the modern techniques which were learned in the Iraq War of cutting the skull open to let the brain have room to swell. Consequently, I went into a coma, where I remained for about twelve hours.

Fearing that they might induce a permanent coma if they administered anesthesia to perform surgery, the doctors postponed my operations until after I had regained consciousness; they waited three days before performing the surgery. Basically, *"If he is still alive on Thursday, we'll see what we can do."*

While I was in the coma, I always had a pulse EKG and an EEG, and so I was never actually pronounced dead. However, a good working definition for the near-death experience is a condition in which "if left alone, the patient would have died." So, I was in the ICU, in critical condition, in a coma, when the NDE took place. Had I not been in the ICU, I would have most certainly died.

To this day, I recall nothing about the accident or arriving in the ER, although the nurses said I was actually coherent on arrival. Instead, my first recollections after the beginning of the accident were when I awoke, out-of-body, on "the other side" in communion with the *Being of Light*. I have no recollection of certain phases of leaving my body as described by other NDE survivors, but I vividly recall returning into the physical body. Surprisingly, not many NDE narrations describe the return to the body except to say, "Suddenly, I was back in my body." Yet I remember the details of this re-entry into physical matter in very clear visceral detail, which I will describe further on.

THE OTHER SIDE OF THE VEIL: The odyssey began for me when I just "woke up" on the other side, with the *Being of Light.* The feeling was as if I had always been there. I felt completely home at last. *The Being of Light* was an old friend who had known me for eons. Time itself had ceased; there was no schedule and no hurry. Dimensional space (3-D as we know it here in physical reality) was not a concept in effect there. (*Years later, after string theory and dark energy had been developed, I would realize that I had been in the additional dimensions of dark energy.*)

But at the time, I just realized that borders and boundaries did not seem to exist in any concrete way. I felt loved, at peace, and as if I was being held like a babe in arms. Yet I knew from deep within me that I was connected to, and an integral part of, the *Being of Light* Itself. We were not separate beings. Instead, I was one essence with the brilliant golden white *light.* Everything was more than okay. I knew that I had been there for quite a while prior to waking up. I was aware that I had been lovingly nurtured, and I awoke feeling "restored."

Now, slowly I became aware that I must return to physical life in time and space (i.e., move back into the physical realm of 3-D plus time). This idea, that I must go back into matter and a physical body, came to me not in words but intuitively, from the *Being of Light.* What I received was a distinct impression (ESP) that communicated:
> *"Yet a little while, and you can return to the Light, but for now you must return to the physical life."*

This awareness, that things were not yet finished with my prior physical life on earth, was the natural result of the things we (the *Being of Light* and I) had been reviewing just before my full awakening to my presence with the *Being of Light.* I was aware that this return to physical life would only be short term and I would soon be able to return home again to the Light. So far, this short-term assignment has been over forty-six years, but still it remains for me only a temporary assignment before I will return to my true home with the *Being of Light.*

I noticed that the *Being of Light*, which I knew was connected with the Supreme Consciousness of the Universe, did not bother to identify itself as being Jesus, God, nor any other name. I just knew that while I was there, the *Being of Light* was part of me and of all things, and that we knew each other intimately, so there was no need for introductions or identity descriptions.

We were part of each other, always had been, and what could be more simple? I had never experienced such a state of connectedness or love. Many years later in my research I would find the clearest description of this state of being a part of everything in the universe in the following passage taken from the Gnostic Gospel of Phillip (discovered at Nag Hammadi, Egypt, in 1945 and released to the West in 1977, these texts were written before 390 AD.):

> *"It is not possible for anyone to see anything of the things that actually exist unless he becomes like them. This is not the way with man in the (physical) world: he sees the sun without being a sun; and he sees the heaven and the earth and all other things, but he is not these things. This is quite in keeping with the truth* (in the physical world).
>
> *But* (when) *you saw something of that place* (eternity), *you became those things.* (When) *you saw the Spirit, you became Spirit.* (When) *you saw Christ, you became Christ. You saw the Father,* (and) *you shall become Father. So in this place* (physical world) *you see everything and do not see yourself, but in that place* (eternity) *you do see yourself and what you see you shall become."*[22] Gospel of Phillip (Nag Hammadi Library Codex II Tractate 3 p.61) *Interpretations in parenthesis are the author's.*

This can be summarized as: *You did see yourself (the* **Light***) in that place and what you saw you shall become.*

MY RETURN TO THE BODY: Now, as I headed back to the physical life and reentered the feelings, pain, and suffering of a physical body, the *Being of Light* slowly began to be obscured by tendrils of reddish purple "blood," which began to wash over my view of the *Light*. The *Light* itself had now faded to an orange "sunset sky" behind a tie-dyed batik curtain of red tendrils. Slowly the red-purple tie-dyed streaks closing out the view of the golden-orange sky thickened until the *Light* was entirely obscured by the red-purple veil.

> *Note: I wrote down all these colors in the months following the NDE, yet I would find later in the Tibetan Book of the Dead the exact same colors being described as occurring in reverse order as one leaves the body just after the death of the physical body.*

In a metaphorical sense, I speak of this as a lowering of the "veil of blood and tears," believing that it represented the separation zone between the spiritual existence and the physical reality.

Many years later I would find the following passage in *A Course in Miracles*, which is clearly describing that same veil:

"In the holy instant nothing happens that has not always been. Only the veil that has been drawn across reality is lifted. Nothing has changed. Yet the awareness of changelessness comes swiftly as the veil of time is pushed aside. No one, who has not yet experienced the lifting of the veil, and felt himself drawn irresistibly into the light behind it, can have faith in love without fear. Yet the Holy Spirit gives you this faith." [23] A Course in Miracles, Chapter 5, Section VI

Being left alone again without the *Light* was now uncomfortable, and *I truly did not want to come back here to this physical life.*

However, subjugating my "will" into the "blood" of this temporal existence, I did come back into the flesh. I reoccupied my body and this process of re-entry caused excruciating pain. I believe this pain was not from my injuries, but was entirely due to the conversion from spirit to physical. It was painful as I "slammed" back into the confinement of a body holding feelings. I could feel the dense heaviness of the body's physical mass, and the pain of having nerves, feelings, and emotions again, the literal thickness of the blood of existence, the hulking, dense mass of physicality. These tangible feelings and the associated pain were all absent while I was with the *Light*; only love and knowledge remained with me there. Communing with the *Light*, only "metaphysics" existed, but when I returned to time and space, I re-entered "physics," the polarization of material existence.

During this re-entry, just prior to *slamming* into the physical, I saw myself surrounded by what I call **candle flames, or sparklers of light**. Later I would read P.M.H. Atwater, another near-death survivor, who records that:
"I floated ever so gently back into her body, moving as I went on a layer of large bright sparklers such as those used on the Fourth of July" [24] P.39 Coming Back to Life: The After-Effects of the Near-Death Experience, P.M.H. Atwater

Apparently, from her description, she also saw the same candle flames (sparklers) that I saw. Later in my research I would find that Thoth-Hermes in the ancient Egyptian records had stated that our soul is **sheathed in flames**. I also found that the first- through fourth-century Gnostic Christians, and the Hellenistic Mystery religions prior to them, all called this flame-filled reentry the "Baptism by Fire." It occurred for them after achieving enlightenment through an out-of-body spiritual communion with the *Light*. This "Baptism by Fire," marked the Gnostic level of enlightenment derived from literally "seeing the light."
Note: It isn't surprising that the original Gnostic Christianity often makes more sense to NDE survivors, while current versions of organized Christianity may seem to be weak, controlled, and watered-down misinterpretations.

Also, particle physicist friends sometimes suggest that the flames or sparklers are merely our expanded awareness annihilating back into the void (zero-point field) as we return to the physical. Now, as yet we have no science on this, just speculation.

THE BRAIN SHIFT: It was just like that, lying there at peace in atonement with the *Light* and the universe, with not a care, and then suddenly *"in the twinkling of an eye"* re-entry into the intense pain and feeling of time and space.

Yet I did not come back as the same guy at all. I had undergone an extreme psychological make-over. The change that took place in my personality is described by many near-death survivors as a *brain shift*. When you return to this life, you are still a regular person; you haven't suddenly become a spiritual guru or shaman, although you may become one later. On the other hand, everything in your paradigm has shifted and you possess a rare gnosis that life is eternally continuous.

It is as if your brain has disconnected like an unplugged computer. The brain has temporarily shut down and now the read-only memory (ROM) is still there, but the random access memory, and anything held temporarily on it, has been wiped clean.

You have returned to re-energize the body's circuits (hardware) and begin to reboot the software, but it will never be as it was before. This return to the prior life is actually a second life, and is not at all a part of the first life. For me, personally, it took many years to overcome the expectations of friends and relatives who still saw me as being the prior personality, and whenever I would act differently from their expectations, they would wonder what had come over me. Even to this day, they do not realize that, although I am the same *spirit*, I am a different *personality* than the one that died earlier. It is a better approximation to simply say that I was *reborn to a new life* (reincarnated) in the same body.

> **Note:** *NDE survivors have actually done what most Fundamentalist Christians claim to have done spiritually. The NDEr has been "born again" in the same body. It is a better approximation to simply say that I was reincarnated a second time into the same body. Also, this is obviously what both Jesus and Saint Paul were actually talking about, where the old person has died and a new person is living there instead.*

The previous life was over, and I now had a new life. The medical professionals may say that I revived, or had never even died. But my memories of that prior life back in B.C. (Before Crash) are somewhat dim, and to describe them I like to say that they are recorded in black and white. To recall those memories of that first life, I have to go search them up in old files, which then have to be unzipped before I can access those dusty ROMs. On the other hand, everything that has happened in this second life is held in RAM (random access memory) and is recalled immediately in vivid living color.

The altered worldview and the psychological changes that also came back with me are listed below:

1. **I have no fear of dying.** While I do fear being injured or broken, I simply have no fear of death itself. I know that death is only a transition and I am fully aware that our consciousness survives that change. I have only returned here again temporarily, and so I am fully aware I can return to the *Light.* I also know that to my *home-with-the-Light* is where I will go when I leave the physical existence.

2. **Material things have almost completely lost their importance to me.** I am still a very responsible worker who pays his bills and maintains good credit with the banks. But I am much more motivated by spiritual things rather than by materialism or success.

3. **I am much more feeling, psychic, able to read other's feelings and express love easily.** I have continually observed many psychic experiences merely because of my changed perspective.

4. **I realize all the world's religions don't have it right when it comes to dying.** But also that each is founded on the honorable principle of attempting to find the truth.

5. **I no longer see any need for individual souls.** I felt a greater empathy for all souls. Later, I revisited this impression and realized that I actually see only one soul, which manifests itself in the physical world in many different egos. Spiritually we are all part of one connected soul, although the consciousness seems to individuate on both sides of the veil.

6. **I no longer believe in retribution after death,** nor a heaven or hell as taught by most Western religions. I understood that there is only something to be learned in this physical life, and that any judgment we make after death, during our life review, is that "it was all good."

7. **I also knew that no one leaves this physical life until their time,** which is not measured as a date, but as a condition of learning. They go when they are finished with learning. Not one second early, nor one second late.

THE TRAUMATIC AFTERMATH:

THE DIFFICULTY OF EXPLAINING ALL THIS: Waking up and finding myself back in the physical world, I immediately wanted to share what I had discovered about the afterlife and consciousness survival, and especially share with others how free that gnosis made me.

> *I had been gifted by the universe with the most precious piece of knowledge mankind, here in the physical plane, can ever have; the total assurance that life continues after physical death. But I quickly found that nobody wanted to hear about it! In fact, they found it disturbing.*

I knew that eternal life was real, that there was no hell to pay, and that our consciousness survived in an alternative dimension. *Oh, my God! What wonderful news is this!* I also knew that eternal life was not something earned down at the church, but was a *free gift for everyone*, precisely as St. Paul had declared in Ephesians:

> *"For it is by Grace you have been saved … it is the gift of God, not of works so that no one can boast.*[25] St. Paul's letter to the Ephesians 2:6-9 (NIV)

THE MEDICAL COMMUNITY DID NOT WANT TO HEAR ME: Immediately after coming back here, when I began to describe my experience, the nurses quickly hushed me up on this subject. They feared for my safety. This was back in 1970, five years before Dr. Moody's book and the near-death experience began to reach the mainstream of medicine. At the time, many nurses knew about this phenomenon, but doctors trained in *Materialist* dogma were still refusing to think about it. Doctors who refused to consider it as real believed the NDE was merely a hallucination. Psychiatrists wanted to condemn it as a psychotic disorder and were anxious to commit to an asylum anyone who persisted in talking about it. So, I quickly learned to keep quiet about it while in the hospital. This is when I learned that the medical establishment, just like the priests, only accept empirical evidence, or scientific evidence *if it happens to agree with what they already wanted to believe.*

Note: I met an NDE survivor in 2014 who had her NDE in 2004 and was placed in the Mayo Clinic until she stopped talking about "that craziness." So, there are still medical doctors out there, in practice, who will commit people to the nuthouse for not agreeing with their chosen version of reality. This travesty of *Materialist* doctors denying religious freedom to others while forcing them to agree with their own religion of *Materialism* continues in the medical community even today.

MINISTERS REFUSED TO HEAR IT ALSO: But after leaving the hospital, I hoped to discuss my experience with religious leaders, believing that would be a safer context and that the clergy would listen to me without trying to commit me to the asylum. But when I spoke with them, they also told me it was merely a hallucination or a dream and simply could not be true **because it disagreed with their interpretation of scripture.**

I was shocked by their desire to remain ignorant on the subject. Here, I had the proof of the afterlife that the church had been seeking for all time. Yet because my specific details were slightly different from their chosen ancient speculations, they wanted to reject it, out of hand. This is when I learned that the churches, just like the medical doctors, only accept empirical evidence or scientific evidence **if it happens to agree with what they already choose to believe.**

My family also had great difficulties with my story. Out of loyalty to me, they wanted to believe me, but found it hard to step away from what they had believed for so long.

THE "STANDARD" NEAR-DEATH EXPERIENCE:

In 1975, five years after I had my NDE, Dr. Raymond Moody, M.D. published the results of his study of NDEs in a book entitled *Life After Life*,[26] which began to give credence to what I had experienced years before. Dr. Moody had studied hundreds of near-death experiences and analyzed the collated results. In that text he described what he had found to be the "standard" near-death experience. He noted that most experiencers did not recall the whole sequence, but all had the majority of the parts. Here is what Dr. Moody reported:

A man is dying, and as he reaches the point of greatest physical distress, he hears himself pronounced dead by his doctor. He begins to hear an uncomfortable noise, a loud ringing or buzzing, and at the same time feels himself moving very rapidly through a long dark tunnel. After this, he suddenly finds himself outside of his own physical body, but still in the immediate physical environment, and he sees his own body from a distance, as though he is a spectator. He watches the resuscitation attempt from this unusual vantage point and is in a state of emotional upheaval.

*After a while, he collects himself and becomes more accustomed to his odd condition. He notices that he still has a "body," but one of a very different nature and with very different powers from the physical body he has left behind. Soon other things begin to happen. Others come to meet and to help him. He glimpses the spirits of relatives and friends who have already died, and a loving warm spirit of a kind he has never encountered before, **a being of light appears before him. This being asks him a question, nonverbally,** to make him evaluate his life and helps him along by showing him a panoramic, instantaneous playback of the major events of his life. (This is the life review or **weighing of the heart.**) At some point he finds himself approaching some sort of barrier or border apparently representing the limit between earthly life and the next life. **Yet he finds that he must go*** *back to the earth, that the time for his death has not yet come. At this point he resists, for by now he is taken up with his experience in the after-life and does not want to return.* He is overwhelmed by intensive feelings of joy, love and peace. Despite his attitude, though, he somehow reunites with his physical body and lives.

Later, he tries to tell others, but he has trouble doing so. In the first place, he can find no human words adequate to describe these unearthly episodes. He also finds that others scoff, **so he stops telling other people.** Still, the experience affects his life profoundly, especially his views about death and his relationship to life[27] *(annotations in parentheses are author's).*

THE LIFE REVIEW: The life review, which is a standard element that occurs in many near-death experiences, is uncannily similar to *the weighing of the heart* depicted in the ancient Egyptian papyrus of *La Pesée du Cœur* (see Appendix A). And during many NDEs, this is the point at which the consciousness makes the decision to return to life after reviewing the emotions of their own heart. This event in the modern NDE agrees with the ancient Egyptian myth where, if your heart is not lighter than a feather, but is still burdened with the cares of this life, then you must return to the physical life. Following are two descriptions of the life review from modern NDE experiencers interviewed by Kenneth Ring, as reported in his 1984 book, *Heading Toward Omega.*

1971 AD—BELLE'S NDE

You are shown your life—and **you do the judging**. Had you done what you should do? You think. "Oh, I gave six dollars to someone that didn't have much and that was great of me." That didn't mean a thing. It's the little things—maybe a hurt child that you helped or just to stop and say hello to a shut-in. Those are the things that are most important... You are judging yourself. You have been forgiven all your sins, but are you able to forgive yourself for not doing all the things you should have done and some little cheaty things that maybe you've done in life? **Can you forgive yourself? This is the judgment.**

1971 AD—DARRYL'S NDE

As the light came toward me, it came to be a person—yet it wasn't a person. It was a being that radiated. And inside this radiant luminous light which had a silver tint to it—white, with a silver tint—was what looked to be a man... Now, I didn't know exactly who it was, you know, but it was the first person that showed up and I had this feeling that the closer this light got to me the more awesome and pure this love—this feeling that I would call love... And this person said, "Do you know where you are?" I never got a chance to answer that question, for all of a sudden, my life passed before me. But it was not my life that passed before me nor was it a three-dimen-

sional caricature of the events in my life. **What occurred was every emotion I have ever felt in my life,** *I felt. And my eyes were showing me the basis of how that emotion affected my life. What my life had done so far to affect other people's lives using the feeling of pure love that was surrounding me as the point of comparison...* **Looking at yourself from the point of how much love you have spread to other people is devastating. You will never get over it. I am six years away from that day (the day of the NDE) and I am not over it yet.**

RESEARCHING IT FOR MYSELF:

Lacking any support whatsoever from the medical community or the mainstream churches, I began to do my own research on what would later become my life's work of investigating the science of the afterlife. Eventually, I found St. Paul's description of his own near-death experience in 2 Corinthians, 12 (as quoted below). Here St. Paul describes his own NDE, which occurred during the decade after Jesus' death (about 35 AD). St. Paul is speaking about his own conversion experience on the road to Damascus, and when he saw the light, but doing so in the third person! He clearly is speaking about an *out-of-body* experience. When the same story is related by Luke in *The Book of Acts*, the being of light that St. Paul is supposedly speaking to on the road to Damascus is identified by Luke as Jesus, but, in St. Paul's own telling of the story in 2 Corinthians, Paul does not identify a "Jesus."

> *"I must go on boasting. Although there is nothing to be gained, I will go on to visions and revelations from the Lord. I know a man in Christ, who fourteen years ago was caught up to the third heaven. Whether it was in the body or out of the body, I do not know—God knows. And I know that this man, whether in the body or apart from the body I do not know, but God knows, was caught up to paradise. He heard inexpressible things, things that man is not permitted to tell. I will boast about a man like that, but I will not boast about myself, except about my weaknesses. Even if I should choose to boast, I would not be a fool, because I would be speaking the truth. But I refrain, so no one will think more of me than is warranted by what I do or say."* St Paul's second letter to Corinthians 12:1-5 (NIV)

Saint Paul describes that he went out-of-body and "heard inexpressible things" just as happens in the modern NDE narratives where the experiencer travels out-of-body to commune with *the Light,* and finds it to be ineffable. And for me, this was concrete evidence that I was not alone. Now I knew that this revered apostle St. Paul, a great Christian, had also "been there" with *the Light,* just as I had. This changed my opinion of him and I then began to study the *real* St. Paul. No longer would I study him from the *proper* perspective of the Roman Church and its prodigal child, the Protestant Church, but instead from the perspective of a fellow near-death survivor.

FINDING TOOLS FOR LIVING IN THIS WORLD: Now, with a new understanding of St. Paul, I searched further in the scriptures and found the following statement in 1 Corinthians, which gave me a tool that I now use to find direction in life.

> *"My message and my preaching were not with wise and persuasive words, but with a **demonstration of the Spirit's power, so that your faith might not rest on men's wisdom,** but on God's power."*[28] 1 Corinthians 2:4-5 (NIV)

Here Saint Paul was describing a tool of the faith that all Christian believers should use in their own lives. St. Paul was clearly telling me, and all Christian believers, that:

> ***"We should not rely on the words of men and their interpretation."***

Instead St. Paul wants us to:

> ***"Rely only on what the Spirit reveals to us personally in our own heart."***

Saint Paul, right there in the NIV was advising us to believe only the gnosis of our own hearts and not what some learned church leader is telling us, if it goes against our own heart. Obviously, Paul was a Gnostic Christian rather than the limited type of believer that later became the only allowable "orthodoxy" on punishment of death.

Today this *"demonstration of the Spirit's power"* St. Paul spoke of in First Corinthians is precisely what I do when I **"demonstrate"** as an evidential medium by delivering "Spirit greetings" from discarnate departed loved ones at Spiritualist churches, bringing through their names and relationships for total strangers.

CONVICTION THAT THE CONSCIOUSNESS SURVIVES:

All NDE survivors seem to have arrived at the same conclusion, that the individual consciousness definitely survives death, and this brings them great peace. This is the gnosis in their hearts.

1958 AD—NDE OF CARROL PARRISH-HARRA: This woman makes the clearest statement of what each NDE survivor believes about death.

> *"Being in that magnificent presence, I understood it all. I realized that consciousness is life. We will live in and through much, but **this consciousness we know that is behind our personality will continue.** I knew now that the purpose of life does not depend on me; it has its own purpose. I realized that the flow of it will continue even as I will continue. New serenity entered my being."*[29]

Today, fifty years later the NDE experience still delivers the same conclusions. Here is what an M.D., neurosurgeon, and Harvard Medical professor found during his own NDE in 2008.

2008 AD—A NEUROSURGEON HAS A NEAR-DEATH EXPERIENCE: In November 2008, Dr. Eben Alexander III, a renowned neurosurgeon and Harvard professor who had spent fifty-four years believing in the generally accepted materialist scientific worldview, had a near-death experience. Previously, Dr. Alexander thought that as a neurosurgeon, he knew how the brain and mind worked. But after this experience in which he spent a week in a coma, he changed his viewpoint completely.

> *NOTE: Dr. Alexander suffered a prolonged case of severe bacterial meningitis. This disease, because of its selective destruction of the outer surface (or neocortex) of the brain, is normally terminal.*

Shocked by the hyper-reality he experienced, which many had reported in NDEs, Dr. Alexander spent the following two and a half years reconciling his experience with contemporary physics and cosmology. He finds that his spiritual experience (or NDE) is *totally consistent with the leading edges of scientific understanding today.* Dr. Alexander believes that, taken together, science and spirituality will thrive in a new symbiosis, offering the most profound insight into fundamental Truth, and I believe he is entirely correct.

But in order for this profound advancement in our culture to happen, many in both the scientific and religious communities must denounce their addiction to prejudiced, closed-minded, dogmatic beliefs (superstitions) and instead allow society to open its awareness to this new synthesis of understanding. Dr. Alexander believes that the progression of individual conscious awakening will cause this movement in society.

> *"One thing that we will have to let go of is this kind of addiction to simplistic, primitive reductive materialism because there's really no way that I can see a reductive materialist model coming remotely in the right ballpark to explain what we really know about consciousness now.*

> *"Having been through my coma, I can tell you that's exactly wrong and that in fact the mind and consciousness are independent of the brain. It's very hard to explain that, certainly if you're limiting yourself to that reductive materialist view."*

- • **7 Network. Sunday Night. Proof of Heaven. Eben Alexander's touch with heaven**.
  When Dr. Eben Alexander collapsed with meningitis four years ago, he spent seven days in a coma. When he finally woke up, he told an extraordinary tale of what he experienced while he was sleeping. His incredible story taps into the age-old question,

debated over thousands of years and by hundreds of cultures: is there an afterlife, and if so, what is it like?
https://youtu.be/BtFZxdRiSek

ANCIENT NEAR-DEATH ACCOUNTS

None of this is new science; the history of near-death experiences in the recorded literature goes back several thousand years throughout the surviving literature of the Ancients, back to at least 1000 BC.

Approx. 1000 BC—JOB:

This narrative is related by the writer of the Book of Job, which most scholars believe was written sometime after the reign of Solomon and prior to the first exile of the Hebrews (Babylonian Captivity). This means it could have been written as early as 931 BC and as late as 597 BC. But the events in Job's lifetime took place sometime between 2000 and 1000 BC, and most likely nearer to 1000 BC, having been transmitted *orally* prior to being written (which of course allows for great distortions to have inadvertently crept in). Job's account of his NDE reads very differently in several ancient Hebrew texts than it does in the current edition of the New International Version (NIV) of the Protestant Bible.

The following verses from the NIV version of Job are a favorite of mine and also many other Christians. Often quoted on Easter Sunday, they come from the Hebrew of the Old Testament.

> **"I know that my Redeemer lives, and that in the end he will** *stand upon the earth.* **And after my skin has been destroyed, yet,** *in* **my flesh I will see God. I myself will see him with my own eyes—I and not another—How my heart yearns within me."** Job 19:25-27 (NIV—text as printed)

But several ancient Hebrew texts for these same verses provide alternative translations. Naturally, the NIV translators have chosen those translations that fit their chosen dogma of resurrection of the physical body. Yet when I examined the alternative texts, which the NIV mentions in the footnotes, I found that those alternative versions actually agree in all details with what NDE has been reporting for thousands of years. And I found the alternative translations contain no resurrection of the physical body but only resurrection of the *spirit.* Here is how the alternative Hebrew text reads:

> *"I know that my Redeemer lives, and that in the end he will stand upon my grave. After I awake, though this body has been destroyed, then (while) yet apart from my flesh (i.e., out-of-body), I will see God. I myself will see him with my own eyes—I and not another—How my heart yearns within me."* Job 19:25-27 (NIV - Alternative Hebrew texts, as listed in NIV footnotes)

This alternative translation shows clearly that Job is describing what he learned during an NDE. Certainly, all the conditions for an NDE were present for him. Earlier in the book, Job's aliments are described.

- He is very sick.
- His entire body is covered with boils.
- He has wasted away to skin and bones.

In such condition he might as well be dead. So, while the details of his disease are sketchy, it is obvious that he has come very close to death (i.e., near-death proximity) from this disease. Knowing that an NDE was probable, and seeing that what Job relates in the above verses 25-27 is the same gnosis near-death survivors return with, it appears most probable that Job had an actual NDE and achieved the gnosis which derives from it prior to relating these verses. This is evident from the several "standard" items of the NDE gnosis he shares here:

1. He knows that after death he will be *out-of-body...*"yet, apart from my flesh."
2. He knows his body will be destroyed, but he will still live on without it.
3. He describes the experience much as it has been described in the NDE reports, "apart from my flesh I will see God, I myself will see God with my own eyes."
4. He knows he will see the *Light* (see God).
5. He knows that he has eternal life and cannot wait to get back there.

Further, Job's remark, **"If a man die, shall he live again? All the days of my appointed time will I wait, till my change come"** (Job 14:14, NIV), reminds me of my own NDE impression that *"yet a little while, and you can return to the Light."*

Approx. 400 BC—Er: Related by Plato (428/427 BC—348/347 BC)
THE NEAR-DEATH EXPERIENCE OF THE GREEK WARRIOR ER: Here is a report of a near-death experience that took place some 430 years before Saint Paul's NDE. This report is taken from Plato's *Republic*, Vol. X, 614b, but this version is my own shortened paraphrase based on several descriptions appearing in multiple translations of Plato.[30]

> *Er was a Greek warrior, the son of Armenius of Pamphylia, who was slain on the battlefield. His countrymen collected the corpses, and hauled them home in carts about the tenth day after the battle. Many of the bodies were already decayed. But his body was still intact and was laid with the others on the funeral pyre to be burned about two days later, about twelve days after the battle. It was there, after being coun ted dead for twelve days, that Er came back to this life. Er then described what he had seen in the realms beyond.*

*First, he said that his soul left his body (out-of-body). Next, he traveled with many other spirits to a place where there were two openings leading from earth into the afterlife. Between these openings sat some sort of Divine Beings. Apparently, the Divine Beings could see at a glance in some sort of display all the things that the soul had done while in its earthly life (**i.e., life review - weighing of the heart**). After every judgment, the Divine Beings told the righteous to go to the right and then upward to heaven. As they did this they attached a token to the souls representing the judgment that had been passed about their life review (**Ankh, or Key of Life**). Others were directed to the left and then downward (**directed to travel through the underworld**).*

All of what Er described is represented in the ancient Egyptian weighing of the heart, when the soul with a heavy heart is directed to the underworld to be led by Anuket to later be born again in a new body, and the righteous soul is given the Key of Life and directed to ascend into the heavens (see Chapter 7).

*However, the Divine Beings told Er that **he was to return and inform men** in the physical world about what the other world was like.*

*NOTE: All modern near-death survivors remember being told that it was **not yet their time**, and that **they must return to accomplish unfinished business**.*

Most near-death experiencers interpret what they see in the perspective of their own religious cultural upbringing; Er was no doubt a Platonist who died about four hundred years before the Christian era. What we have in Plato's narrative is Er's interpretation of what he saw colored in the light of his prior religious beliefs. Regardless of that, the reported experience was almost exactly the same as every other reported NDE throughout history right up to today.

NDE research has shown that everyone—myself, Saint Paul, Er, and all other near-death experiencers from many different religions and cultures—all report the same events. They may interpret its meaning differently depending on their different religious backgrounds, but they all report some or all of the same basic events, as itemized in this list taken from, P.M.H. Atwater's 1988 book, *Coming Back to Life*:

1. A feeling of floating out of their body
2. Passing through a darkness or tunnel
3. Ascending toward a light at the end of the tunnel
4. Greeted by friendly voices, people, or beings
5. The life review, or seeing a panoramic view of the life just lived, with assessment of the meaning of the events (*the weighing of the heart*)

6. A reluctance to return to the earth plane
7. Warped sense of time and space while over there
8. Disappointment at being revived

For further study on this subject, here are several YouTube videos. The first is a September 2013 television interview in which I describe my NDE and also my 2012 book, *The Death Experience: What It Is Like When You Die*. The second is an hour-long film that collates the experience of myself and four other near-death experiencers, with commentary by Dr. Dean Radin, chief scientist at the Institute of Noetic Sciences, Consciousness Research Laboratory. The third is a three-minute blurb from the Institute of Noetic Sciences describing what they do, by Dr. Radin.

- **"SCIENCE OF THE AFTERLIFE,"** Dr. Alan Hugenot. What it is like when you die. http://www.youtube.com/watch?v= sG8RAVh4VwE
- **"BEYOND OUR SIGHT"** Dr. Alan Hugenot discusses the near-death experience and mediumship, along with Dr. Dean Radin and four other mediums and NDE survivors (shown at IANDS 2014, 57 minutes) https://www.you tube.com/watch?v=xpSuO8DtiMM
- **Dr. Dean Radin, chief scientist at Institute of Noetic Sciences** (noetic.org) discusses consciousness studies at Noetic.org VIDEO CLIP (2 minutes) at http://vimeo.com/113981492
- **7 Network. Sunday Night. Proof of Heaven. Eben Alexander's touch with heaven**.

When Dr. Eben Alexander collapsed with meningitis four years ago, he spent seven days in a coma. When he finally woke up, he told an extraordinary tale of what he experienced while he was sleeping. His incredible story taps into the age-old question, debated over thousands of years and by hundreds of cultures: is there an afterlife, and if so, what is it like? https://youtube/BtFZxdRiSek

CHAPTER 7

SCIENCE OF CHILDREN
WHO REMEMBER PAST LIVES:

"There are three claims in the ESP field which, in my opinion, deserve serious study... One of these is that **young children sometimes report the details of a previous life, which upon checking turn out to be accurate...** *and which they could not have known about in any other way than reincarnation..."* —the late Dr. Carl Sagan

"The transmigrations (reincarnation) of souls was taught for a long time among the early Christians as an esoteric and traditional doctrine which was to be divulged to only a small number of the elect." —Jerome (340-420 AD) in a letter to Dimetries

SCIENCE AND REINCARNATION: There is much rigorous Western science that fully supports reincarnation. As with NDE, we again have enough data that proves reincarnation is also a scientific fact. The only reason the scientific establishment resists it is because it refutes their own unrecognized religion.

At the University of Virginia in Charlottesville, the late Ian Stevenson, M.D. (1918–2007), former Carlson professor of psychiatry, studied this business of "taking life up again" for over fifty years, from the early 1960s until well into this first decade of the twenty-first century. But before we look at his magnificent work, first let's look at some history.

WHERE DID THIS SINGLE-LIFETIME IDEA COME FROM? In the West, according to a spring 2007 poll, 23 percent of people over fifty years of age believed in reincarnation. An earlier poll found that 25 percent of American adults eighteen and over believed in reincarnation, 20 percent just "did not know," and 54 percent "knew it wasn't true."

But the majority opinion *in the world* is that reincarnation is a known fact. Here is what Dr. Ian Stevenson found:

"That's the paradox... In the West people say 'Why are you spending money to study reincarnation when we know it is impossible?'...In the East people say 'Why are you spending money to study reincarnation when we know it is a fact?'"

At first, the idea of multiple lifetimes might seem strange to anyone raised in Western society. But a quick study of demographics reveals that the *single-lifetime theory* is only found in populations descended from the Roman Empire. Hey, there is a clue... *Why is this single-lifetime idea only found in the West and Mideast?*

Digging a little further we find that before the Roman emperors took control of the Church and began to rewrite its beliefs with creeds and dogmas, the early Christians, like the bishop at Alexandria in 200 AD, named Origen, definitely believed in reincarnation.

"Every soul comes into this world strengthened by the victories or weakened by the defeats of his previous life. Its place in this world as a vessel appointed to honor or dishonor is determined by its previous merits of demerits. Its work in this world determines its place in the world which is to follow this one." —Origen (185-253 AD)

Even the Nicene Creed (allegedly from 325 AD) ends with these words:

*"I look for the resurrection of the dead and the **LIFE OF THE WORLD TO COME.**"*

Now, just on the surface, that *"resurrection of the dead and life of the world to come"* certainly sounds a lot more like Origen's reincarnation into the *"world which is to follow this one"* than it could ever sound like the claim of the Western Christian churches that it means *"the heaven to come..."*

Honestly, looking at simple semantics, if the Christian bishops at Nicaea actually meant *"heaven to come"* then why didn't they just say *"heaven to come"*? Why in the world (excuse the pun) would they say "world" when they really meant "heaven"?

A little more comprehensive study of history finds that it was actually Emperor Justinian, not the Pope, who convened an Ecumenical Council in 553 AD and got rid of reincarnation. This fifth Ecumenical Council of the Church (Second Council of Constantinople) was a travesty. Convened 550 years after Jesus by the emperor and not the Pope, this was entirely a political convention. Indeed, the sitting Pope Vigilus would not even attend, although records show that he was in Constantinople at the time. He saw that Justinian only invited bishops who believed in the single-lifetime theory and had refused to invite any bishops who believed in reincarnation.

After this allegedly "ecumenical" council passed a resolution to shun all those who believed in reincarnation (in other words all the bishops who weren't there), Emperor Justinian used the Roman Army to enforce this new single-lifetime theory. But not surprisingly in the official records of the ecumenical council, there is no mention of passing this resolution. Some historians believe that Justinian faked the entire thing. He needed only one lifetime so people would have to kiss up to the church.

It is pretty obvious then to any unbiased biblical historian that this was a politically motivated change in the religion. Reincarnation gives people second chances to continue to grow spiritually, and so Justinian needed people to believe in *only one lifetime* in which to get our soul right with God so that he could use the Church and the threat of eternal damnation in hell to control the people. So, for strictly political reasons, the emperor, not the Pope, got rid of the truth of reincarnation.

Twentieth-century biblical scholarship on early Christianity has conclusively shown that reincarnation was the accepted doctrine in early Christianity. It is now well documented that reincarnation was taught by the early Christian fathers, including Origen of Alexandria (185–254 AD), who is considered one of the greatest of the early Christian theologians.

> *NOTE: This means that the theology of the early Christian church of the first century, which included reincarnation in their original Christian Salvation Myth, was IDENTICAL to the ancient Egyptian Salvation Myth from 2,450 years earlier.*

Further, St. Augustine himself confirms that the Christian religion of biblical times was a reawakening of ancient enlightenment of Thoth-Hermes in the Hebrew community much as it was already being practiced in the Hellenistic Mystery religions throughout the known world for at least 500 years before Christ[31].

Christian Bishop St. Augustine (November 13, 354 AD—August 28, 430 AD): in 427 AD wrote "the Retractions" three years before his death, in order to set straight things he had said that were misinterpreted by the Vatican, and in his Retractions he said that this so-called "new" Christian religion was merely a rebranding of the pre-existing "pagan" religion:

> *"That which is called the 'Christian' religion existed among the ancients, and never did not exist, from the beginning of the human race until Christ came in the flesh, at which time the true religion WHICH ALREADY EXISTED began to be called Christianity."*[32] St. Augustine, *Retractions.* Taken from P.54 *The*

Fathers of the Church, Saint Augustine, The Retractions. Brogan, Catholic University of America Press

Augustine's Retractions were written as a clarification in 428 AD, after publication of his seminal work, *City of God*, in 427 AD.[2] My copy of this statement is authenticated by the Roman Catholic Church that this is precisely what St. Augustine said (see the end note).

My own copy of St. Augustine's text, published by the Catholic University of America Press, caries the Nihil Obstat of John C. Selner, S.S., S.T.D. the Censor Librorum, and also the Imprimatur of Patrick Cardinal A. O'Boyle, D.D. Archbishop of Washington.

These are both official declarations that a book or pamphlet is **FREE OF DOCTRINAL OR MORAL ERROR.** *No implication is contained therein that those who have granted the Nihil Obstat and Imprimatur agree with the contents, opinions or statements expressed. In laymen's terms, this means* **that the Catholic Church officially agrees that these are the actual words of Saint Augustine himself, and they agree that he actually said,** *"That which is called the Christian religion existed among the ancients, and never did not exist, from the beginning of the human race." Saint Augustine clearly admits that paganism was in fact the Christian religion. Consequently, seeing the Roman emperor Justinian condemn paganism (Origenism) 120 years later illustrates that the emperors had no desire for true religion, only power and control. That original ancient religion St. Augustine identified as the true faith included reincarnation and allowed for second chances without hellfire.*

Given the above-documented historical facts, honest science causes one to ask, **"How did the original salvation myth get so grievously altered?"** That history is fully explained in Appendix A.

Of course, this well-proven history cuts too close to the bone for most Christians, both Catholics and Protestants, leaving them with the *OMG! We-have-been-lied-to* feeling.

Shortly after hearing me describe this, most "true believers" decide that I (the messenger) am the liar, and set out to "kill the infidel" so that their *believed truths* (lies) can be left intact. So, we'll just let their beloved lies lie, and instead look at the scientific data regarding reincarnation.

1. THE WORK OF DR. IAN STEVENSON:

Dr. Stevenson carried out lengthy investigations of over 2,400 cases of children remembering past lives, in a study that took place over fifty years, working in Eastern and Western cultures. And Dr. Stevenson's colleagues, Bruce

Greyson M.D., Jim Tucker, M.D., et al., are continuing his work, which has, to date, cataloged over 2,500 individual cases of children who remember their past lives. This research has all been carefully documented using rigorous science to weigh the evidence in Stevenson's books: *Twenty Cases Suggestive of Re-incarnation*[33] and *Children Who Remember Previous Lives*[34]. Jim B. Tucker M.D. also published, in 2005, on this same subject his book *Life Before Life*[35].

Stevenson and his department meticulously investigated thousands of cases in which children between three and five years old began relating memories of an earlier life to their parents in the present life. These children often have detailed accounts of their wives, husbands, and children from the previous life.

Often the children have graphic memories of how they died and/or who killed them. They can even recognize former friends and family members still living when they return to the vicinity where the previous life took place. In a few cases, they can also remember and replicate the dialect of the prior home. In other cases, they carry a physical birthmark or physical deformity that corresponds with an injury received in the previous life. Stevenson published another book, *Where Reincarnation and Biology Intersect*, which shows photographically these corresponding deformities that match the prior incarnation.

Yet even by the time these children reach age eight to ten, they usually begin forgetting the prior physical life as they assimilate into their current physical life. If they are surrounded by a culture that encourages these types of prior-life memories, then the memories may continue well into adulthood.

Yet even in the west there has always been an undercurrent in Western culture and literature that acknowledges these memories. Consider the eighteenth-century British poet William Wordsworth.

> *Our birth is but a sleep and a forgetting:*
> *The Soul that rises with us, our life's Star,*
> *Hath had elsewhere its setting,*
> *And cometh from afar:*
> *Not in entire forgetfulness,*
> *And not in utter nakedness,*
> *But trailing clouds of glory do we come*
> *From God, who is our home*

> "Ode: Intimations of Immortality"
> From *Recollections of Early Childhood*
> By William Wordsworth (1770–1850)

Here are a few examples of the cases Dr. Stevenson has verified. (My favorite of all his case studies is this one about Marta Lorenz, for two reasons: it was kept single blind, and it is a great love story.)

1. MARTA LORENZ CASE[36]: This is one of my personal favorites because it reads like a dramatic novella. However, it has been carefully researched by Dr. Ian Stevenson, and also because the return to the second life was "forecast" during the first life. The parents, having been given the prediction before the birth, wisely chose to observe the second life (their daughter) under strict single-blind laboratory conditions.

The results of this decision to blind the "experiment" provided amazing evidence. Here are the details as meticulously documented by Dr. Ian Stevenson, who interviewed all the players except the original Sinha, who died in 1917.

Maria Januaria de Olivero, nicknamed Sinha or Sinhazinha, was born about 1890, in the village of Dom Feliciano, about one hundred miles southwest of Puerto Alegre in the province of Rio Grande de Sul, the westernmost state of Brazil. After Sinha's father disapproved of one of her suitors, the young man committed suicide. Sinha, who loved the young man, was distraught and so exposed herself to the elements and to tuberculosis, acquired an infection of the lungs and larynx, and then, a few months later, she also died.

However, on her deathbed she acknowledged to her dear friend, **Ida Lorenz,** that she wanted to die and had tried to become infected with TB. Then, she also promised Ida that she would return again and be born as Ida's daughter. Sinha further predicted that:

> *"When reborn and at an age when I can speak on the mystery of rebirth in the body of the little girl who will be your daughter, I SHALL RELATE MANY THINGS OF MY PRESENT LIFE, AND THUS YOU WILL RECOGNIZE THE TRUTH."*

Sinha died the day after her declaration to later return as Ida's daughter, in October 1917. At that time she was about twenty-eight years old.

Ten months later, on August 14, 1918, Ida Lorenz gave birth to a daughter, whom she named Marta. Ida and her husband had agreed to maintain the single-blind conditions and carefully never spoke of Sinha's prediction to their other children, nor to the neighbors, or to anyone who might ever come in contact with any future daughters. It was to be the parents' personal secret.

But when Marta was two and a half, she began to speak, exactly as predicted by Sinha, of *when she had been "big before."* She volunteered to her parents that she had been *nicknamed Sinha* in her prior life, and also that her real name in that prior life had been *Maria*.

Marta said all this despite the fact that the names Sinha and Maria had never once been uttered in their household. Marta made no less than *120 separate declarations over the next few years about details of Sinha's life.* All were carefully recorded by her parents. Her home from the former life was only twelve miles away, but her father, in order to keep the single-blind conditions, never took her there until she was twelve years old, although she had asked numerous times to go. And it was not until the time of this visit, at age twelve, that the parents from the prior life heard anything about Marta being the reborn Sinha.

Remember that Dr. Ian Stevenson interviewed *all* the participants, except for the original deceased Sinha, and continued contact with Marta for many years. In the 1974 edition of his book, after his most recent visit with her in Puerto Alegre in 1972, he described Marta at age fifty-four:
> *"She is very much Marta, but still remembers much detail from her former life as Sinha, especially her bout with tuberculosis, and her death. She sees her prior life and this life as different chapters in a continuum."*

When Marta was a child of three or four, and bereaved adults would suffer from grief, Marta would comfort them by saying,
> *"I died, and I am living again."*

And another time, during a rainstorm, when one of her sisters expressed regret that their dead sister Emilia would get wet in the grave, the child Marta said,
> *"Don't say that, Emilia is not in the cemetery. She is in a safer and better place than we are; her soul never can be wet."*

So, although Marta's memories have faded since the vivid days of her childhood, it is apparent from what she said in early childhood that:
> *As a child she also remembered her ten months of life on the other side between the two lives here.*

I highly recommend reading this particular case in Dr. Stevenson's original report, in the book end-noted at the start of this section about Marta. The narrative runs twenty pages and is heavily documented and researched. The interesting story reads like a drama.

2. BISHEN CHAND CASE:[37,38,39] Bishen Chand Kapoor was born into the Gulham family in 1921 in Bareilly, India. At the age of one and a half, he began asking questions about a town called Pilibhit, which was located about fifty miles away, a place no one in his family knew much about and did not know anyone who lived there. Bishen asked his parents to take him there, and it was obvious to them that he believed he had lived there in a prior life. By the time Bishen was five years old, he could clearly articulate memories of his former life and from that village where he and his present family had never visited.

By 1926, he was stating that in the former life he had been named Laxmi Narain, was the son of a wealthy landowner, and remembered an uncle named Har Narain. Later, it would be discovered that this "uncle" was actually the father of Chand's remembered self with that same name of Laxmi. The young Chand described the house of Laxmi Narain, including various features of the layout, as well as a neighbor's house with a green gate, and how he had enjoyed the singing and dancing of the young women who entertained men in bars. He often spoke and could even read words of Urdu, a language written in Arabic script, even though Hindi was the language spoken in Chand's home and no one had taught him Urdu or how to read it.

> **Note:** *These two items are remarkable, a mere five-year-old child remembering how he enjoyed watching the dancing girls in adult bars during his past life, and also speaking and reading a foreign language that he had not been taught.*

A local attorney, K.K.N. Sahay, heard about Chand's memories and came to the house to record statements from the boy and other family members. In one of those most surprising memories, the boy recounted killing a suitor of his mistress, **showing surprising awareness for a five-year-old of the difference between a wife and a mistress**. None of the family nor the attorney had ever heard of the real Laxmi Narain. It is important to know that the attorney wrote down Chand's complete narrative at this time *before the attorney or Chand ever traveled to Pilibhit.*

When the attorney, along with the young Chand and Chand's father, traveled the fifty miles to Pilibhit, it was immediately confirmed that a man named Laxmi Narain had indeed shot and killed a rival lover of a prostitute who was still in the town. Narain had avoided prosecution because of his wealth, but had died two years later at age thirty-two. By the time Chand was five, not quite eight years had elapsed since the death of the adult Laxmi Narain.

When taken to Laxmi's old school, the boy ran to the classroom, described the teacher, and from an old photograph, identified and named classmates, one of whom was in the crowd that had gathered. Chand had a heartwarming reunion with Laxmi's mother, whom he greatly preferred over his own biological mother. The green gate was seen as described, and when given Laxmi's tabla drums, he was reported to have played them with great skill. Before leaving the house, Chand revealed where he had, during the past life as the adult Laxmi Narain, hidden some gold coins, which were recovered the following day.

This story clearly suggests that the consciousness survives in a very real form. The facts of Chand's language and musical skills, the adult emotions this toddler displayed, his identification of people by name from a photograph, even though he had never met them in this life, all are irrefutable proof that the consciousness survives. This case has been carefully verified and documented by the researchers at the University of Virginia. So it would appear that we have at least one WHITE RAVEN.

3. THE SHANTI DEVI CASE:[40] Shanti Devi was born in 1926 in old Delhi. At about the age of three, she began to tell her parents stories regarding a former life, in which she was married to a man named Kedar Nath, who lived in the nearby town of Muttra. In that prior life she also had two children, and she had apparently died in childbirth bearing a third child in 1925. She described in detail her home in Muttra where she had lived with her husband and children.

When her parents tired of trying to stop her from telling these stories, her grand-uncle, Kishen Chand, sent a letter to Muttra to see how much, if any, of the little girl's story was true. He mailed it to an address that Shanti herself remembered and gave to him. That letter was received by a widower named Kedar Nath. His wife, Lugdi, had died in childbirth in 1925, the year before Shanti was born. Quite naturally, Kedar Nath suspected this was all a fraud. So he sent his cousin, Mr. Lal, who lived in Delhi to visit Shanti at her home in Delhi. The cousin would know if she was an imposter.

Mr. Lal went to Shanti's home without announcing his relationship and pretending to be on business. But when Shanti answered the door, she screamed and jumped into his arms. Her mother then came to the door. Before Mr. Lal could speak, the nine-year-old Shanti said,

"Mother, this is a cousin of my husband! He lived not far from us in Muttra and then moved to Delhi. I am so happy to see him. He must come in. I want to know about my husband and my sons."

Mr. Lal confirmed for Shanti's parents all the facts she had told them over the years. Soon they reached the consensus that Kedar and his son should come to Delhi to meet Shanti. Later, when Kedar arrived, Shanti, although only nine years old, treated him as a normal adult wife would do. She kissed him and called him by pet names. Shanti served him biscuits and tea. And then when Kedar's eyes began to fill with tears, she comforted him using personal phrases and endearments known only to Kedar and his late wife, Lugdi Nath.

News reporters decided to sponsor Shanti on a trip to Muttra. They thought that she should be able to lead them to her former home. When the train arrived in Muttra, Shanti saw her relatives from her prior life on the platform (relatives who had not been to Delhi, and whom she had not seen in this life). She yelled and waved at them as the train slowed at the platform. Before getting to meet the relatives, she told the reporters that they were the mother and brother of her husband. On the platform, she began to speak with them. The reporters noted that she had correctly named them. Then, **the reporters realized that nine-year-old Shanti was not speaking Hindustani, the language she spoke at home in Delhi, but was instead speaking coherently in a dialect specific to Muttra.** Shanti had never learned, nor been exposed to, this strange dialect, but according to the news reporters, she spoke it fluently, and if she actually was the deceased Lugdi's reincarnated consciousness, then she would easily know this local language.

Note: this use of the dialect was reported in the news accounts, but could not later be verified by any eyewitnesses for Ian Stevenson during his research some decades later.

Next, as the reporters had requested, Shanti led them to the Nath house. Although she had never been in Muttra during her current life, she led them directly to the Nath house. On the way, she told them things that only Lugdi could have known. Lugdi's husband, Kedar, asked her specifically where she had hidden some rings that she had owned before she died. Shanti told him that they were hidden in a pot she had buried (during her former life) in a location at their prior home where they had lived together before moving to this house. Later, the investigators were able to dig up the rings where she had said they were buried. The records also indicate that everyone who had previously known Lugdi in Muttra now accepted Shanti as being Lugdi's reincarnated consciousness.

Shanti's case has been investigated by numerous researchers over the years. Clearly, cases like this, with remembered foreign language, remembered buried objects, and spontaneously recognized faces from the prior life, are indeed rare. Yet many of them do exist. And just one such case, with so many facts revealed that were unknowable to anyone except the departed's consciousness from the prior life, would seem to be all that is required as compelling proof that reincarnation actually happens. Now, we have a third white raven.

4. THE BONGKUCH PROMSIN CASE:[41] Bongkuch was born February 12, 1962, in the village of Don Kha in the Nakhon Sawan Province of Thailand.

As soon as the child began to speak coherently, he started to repeatedly describe his prior life.

He said that he came from the village of Hua Tanon, which was nine kilometers away, and that his name in this prior life had been Chamrat. He also gave his former parents' names. By the time he was two years old, he told details of his prior family and also that two men had murdered him at a fair in Hua Tanon, stabbing him in several places. He said they had taken his wristwatch and neck chain, and dragged his body into an open field. Bongkuch also said that after his death as Chamrat, he stayed on a tree near the site of the murder for approximately seven years.

Then, one rainy day, he saw his present father passing by and decided to accompany him home on a bus. Bongkuch's father later recalled that he had attended a meeting in Hua Tanon not long before his wife became pregnant with Bongkuch. He also recalled that it had been raining. Bongkuch's parents had never heard of Chamrat's murder before hearing it from Bongkuch.

When Bongkuch was two and a half years old, however, Chamrat's parents and friends heard about what Bongkuch had been saying, and some of them came to visit him in Don Kha. Through these contacts, Chamrat's parents eventually verified everything the young boy was saying about his prior life and the murder that had taken place some ten years before.

Dr. Ian Stevenson later interviewed some of the policemen who investigated the original murder, and they verified that the suspects in the case, one of whom fled, and the other who was acquitted for lack of evidence, were the same two people that Bongkuch had named as being Chamrat's murderers.

Reports of Bongkuch's case appeared in the newspapers in Thailand in March 1965, and the director of the Government Hospital in Nakhon Sawan, Dr. Sophon Nakphairaj, sent copies of those reports to Ian Stevenson, who began studying the case in 1966, and actually met Bongkuch in March 1980.

Bongkuch often spoke Laotian words (Chamrat was Laotian) that his Thai parents did not understand. He also felt himself to be an adult, imprisoned in a child's body. He often made advances toward teenage girls, but would ignore girls his own age. Once, at a very young age, he attempted to fondle a teenage girl who was visiting his parents. Yet by the time he was eighteen, he had nearly completely moved on from those childhood memories. Apparently, this is another white raven.

- **SCIENTIFIC CASES FOR REINCARNATION**: Uploaded on Oct 23, 2011. This interview with Dr. Jim Tucker, MD, current

(2016) director of the Division of Perceptual Studies at the University of Virginia and successor to Dr. Ian Stevenson, describes the Division's ongoing efforts to investigate the phenomenon of spontaneous past-life memories in very young children. https://youtu.be/88s6HHkbh84

2. *REINCARNATION IN THE WEST:*

As Western culture has shaken off the restrictive thinking of Christianity, we have begun to embrace the wider views of the world, and more people have recognized that children are remembering their past lives. Here are some examples.

5. JAMES LEININGER:[42] A two-year-old in Lafayette, Louisiana, remembered his prior life as James Huston and recalled being shot down during World War II on March 3, 1945, while flying his Navy Corsair over Chi Chi Jima. Huston was a pilot flying from the aircraft carrier USS *Natoma Bay CVE-62,* and the child James Leininger remembered Huston's friend, Jack Larson, a fellow pilot on the ship.

Jack Larson is still living. The child remembered that his plane was hit near the engine and crashed in flames. Eyewitnesses of his crash have verified the fact that his plane was hit in the engine. James Huston's sister is still living, and has now come to believe that James Leininger is actually her brother reincarnated.

She believes this simply because of all the things little James knows that only her deceased brother could have known. His parents published their story, *Soul Survivor,* in 2010. Here are two videos that tell his story. The second one is from FOX News

- **REINCARNATION PROOF: THE JAMES LEININGER CASE:** Published on Jul 7, 2013. A young American boy remembers his past life as an American fighter pilot fighting the Japanese during World War II. He recalls information that only his past life sisters and his own past life incarnation know. Meeting up with old pilots, he manages to name them all and they confirm his story. The statistics that he can do the aforementioned are off the scale. We can say clearly, to claim a child's imagination, watching old documentary war films or playing war games on gaming machines are the explanation rather than symptoms, are absolutely ludicrous statements and far removed from reality. To proclaim such is to expose oneself as a pseudo-sceptic, with a belief system in debunking that is on par with cult-like thinking. https://youtu.be/VnXxC-nVsJY

[lb] FAMILY BELIEVES SON IS WORLD WAR II PILOT, REINCAR-NATED - FOX NEWS: Published on Oct 28, 2013. Is this family's incredible story proof of reincarnation? In the book, *Soul Survivor: The Reincarnation of a World War II Fighter Pilot*, Bruce and Andrea Leininger tell the story of their son, James, and the strange nightmares he began having as a toddler. Bruce says the dreams were occurring four or five times a week, and they really began to wonder what was happening when James produced some drawings that were eerily accurate. He was able to draw what clearly looked like a plane being shot down and oddly, he signed his name as "James 3." When his parents asked why he was putting a three after his name, he said he was the "third James." James was also saying the names of WWII fighter pilots and the name of the ship that fighter pilot James Huston served on. Through years of research, the family tracked down the family of James Huston, who was killed in a mission near Japan. They believe all of these inexplicable occurrences point to the spirit of Huston living within their son. https://youtu.be/I9u2EpK35PY

Note: James Leininger is not the child whose stupid and lying parents created the hoax of an NDE in order to sell a book, *The Boy Who Came Back From Heaven,* **and the movie that grossed ninety-two million before they finally admitted their fraud. But such charlatans profiting from the media only hurt the truth of reincarnation and NDEs.**

3. CHILDREN OF THE HOLOCAUST:

Dr. Brian Weiss, M.D. (whose work will be discussed later in Chapter 9), is a psychiatrist who works in past-life regression hypnotherapy. He told me of a Swedish psychologist who has worked with over 200 young people who remember prior lives in the German concentration camps of World War II. That Swedish psychologist has gone back to Poland and other locations of the concentration camps and has validated the accuracy of over two dozen of these children's memories.

For example, *some of these people, who are now thirty years old, read their concentration camp numbers as if they were tattooed on their arms.* Later, when the prior identity associated with that ID number is found in the camp records, it is found that these thirty-year-olds are remembering the proper details of the lives of those concentration camp victims (who had that same ID number), who are now dead.

1974 - Teuvo Koivisto was born into a Lutheran Christian family in Helsinki, Finland, August 20, 1971. At the age of three years, he told his mother that he had been alive before. Teuvo explained that he was told that he was being taken to the bathroom, but he was taken instead to this big "furnace" room.

People were told to undress, and their gold teeth and eyeglasses were taken from them. People were then put in the "furnace" room. He told how gas came out of the walls and he could no longer breathe. Check it out at www.iisis.net/...jewish-**concentration-camp**...change-religion&hl=en_US

1957 - Barbro Karlen was born to Christian parents in Sweden in 1954. When she was less than three years old, Barbro told her parents that her name was not Barbro, but Anne Frank. This was less than ten years after Anne Frank had died in the Bergen-Belsen concentration camp in 1945. Barbro's parents had no knowledge of who Anne Frank might have been. The book, *Anne Frank: Diary of a Young Girl*, also known as *The Diary of Anne Frank*, had not yet been translated or published in Swedish.

Barbro stated how her parents desired that she call them Ma and Pa, but she knew they were not her real parents. Barbro instead said that her "real parents" would soon come to get her and take her away to her "real home." While still a child, Barbro told her parents details of her life as Anne. Barbro also had nightmares as a child, in which men ran up the stairs and kicked in the door to her family's attic hiding place. She told all of this to her parents with no knowledge of the story of Anne Frank. When she was ten they took her to Amsterdam, and she led them to Anne Frank's house and convinced her mother, by what she knew about the place, that she was previously Anne Frank. The most remarkable thing about Barbro (Karlen) Full is how much she looks like Anne Frank. Here is the video URL:

- **ANNE FRANK BARBRO KARLEN FULL:** Published on Aug 28, 2013. Barbro Karlen shares memories from her childhood of being Anne Frank in a past life. Like Anne, Barbro was a child prodigy writer. In adulthood, to overcome her phobia of uniforms and to work with horses, Barbro served as a mounted policewoman for the Swedish national police force. Anne Frank was persecuted as a Jew, whereas Barbro was born into a Christian family, showing that religion can change from lifetime to lifetime. https://youtu.be/PxsC-TGAwsQ

JEWISH BELIEFS ABOUT REINCARNATION: Rabbi Yonassan Gershom, a Hasidic storyteller, teacher, and writer, has published several essays on Jewish spirituality that have appeared in numerous periodicals and anthologies. He has published three books on the subject:

- *Jewish Tales of Reincarnation*[43] is a collection of seventy traditional and modern stories about Jewish reincarnation.
- *Beyond the Ashes*[44] is his account of personal encounters with hundreds of people from all walks of life who have shared their memories of visions, dreams, and flashbacks that seem to be coming from another life during the Nazi Holocaust.

He also presents Jewish teachings about karmic cycles, the levels of the soul, views of the afterlife, and reincarnation in Judaism, as seen in the light of traditional Jewish texts and modern discoveries.

- *From Ashes to Healing*[45] completes the saga begun in the first book by presenting fifteen testimonies of people who have past-life memories, visions, and dreams about the Holocaust.

This all dovetails with many of Stevenson's reincarnation cases investigated in Southeast Asia, where the children specifically remember being Japanese soldiers in their prior lives.

4. CONCLUSION:
So, is reincarnation a scientific fact?
1. We have a logical possibility.
2. We have a great deal of evidence for it.
3. There is no evidence against it.

Therefore, by our scientific requirement for proof beyond a reasonable doubt, **it is already an established scientific fact.** The only thing stopping recognition of reincarnation as a *fact* by the scientific establishment is their own fully unfounded religious superstition.

I recently found this video, which interviews Tom Harper and has clips from Jim Tucker, Anne Frank/Barbro Karlen, and James Leininger/James Huston Jr.
- **EVIDENCE OF REINCARNATION:** An excerpt from *Supernatural Investigator: Reincarnation* by David Brady Productions. https://www.youtube.com/watch?feature=player_embedded&v=9w2MCpzE8u0 ns.

CHAPTER 8

SCIENCE OF AFTER-DEATH COMMUNICATIONS:

"We have no warrant for the assumption that the phantom seen, even though it be somehow caused by a deceased person, is that deceased person, in any ordinary sense of the word."[46] —Frederic W.H. Myers

"I confess that at times I have been tempted to believe that the Creator has eternally intended this department of nature to remain baffling, to prompt our curiosities and hopes and suspicions all in equal measure, so that, although ghosts and raps and messages from spirits are always seeming to exist they can never be fully explained away." —Dr. William James, 1905

"There is a recognized scientific principle that one unknown cannot be used to explain another. We may not understand spirit survival, but do we understand mind reading? According to strict scientific materialistic principles one is just as impossible as the other." —Charles H. Hapwood in *Voices of Spirit*, c.1992

HOW MANY PEOPLE REALLY BELIEVE IN AN AFTERLIFE AND SURVIVING CONSCIOUSNESS?

In the last chapter we looked at the statistics for reincarnation and found that one in four people believe in it. The statistics for belief in the afterlife show that more than half of all Americans believe in some form of consciousness survival. In 2007 *AARP Magazine* reported a survey which found that among the Americans interviewed:

"...more than half of those responding reported a belief in spirits or ghosts—with more women (60%) and men (44%) agreeing that they believed in ghosts. Boomers are more likely to believe in ghosts (64%) when compared with those in their 60's (51%) or 70's or older (38%). Their belief is not entirely based on hearsay evidence either, 38% of all those responding to our poll say they have felt a presence or seen something that they thought might have been a spirit or a ghost."[47] Sep/Oct 2007 *AARP Magazine*

One of the strongest proofs of the afterlife accessible to the average person is an after-death communication (ADC) through mediumship, dreams, or spontaneous apparitions. It provides one's heart with a new form of personal gnosis about the continued existence of a loved one.

One way to prove that human consciousness survives bodily death would be to establish reliable communications under laboratory conditions with those in the afterlife. Scientists who realized this established two societies in the 1880s for this exact purpose. In Great Britain in 1882, the Society for Psychical Research (SPR) was formed, and a sister organization, the American Society for Psychical Research (A-SPR), was formed in 1884. These two organizations attracted a membership of world-class scientists (like William James and Sir Oliver Lodge) and have collected data from hundreds of mediums communicating with disincarnate souls, and documented hundreds of apparitions of the recently dead.

MEDIUMSHIP IS REAL: In 2006, I personally decided, as a scientist, to investigate mediumship. Having read what these scientists had discovered a hundred years ago, I chose to be my own guinea pig and study evidential mediumship myself. Today, after ten years of graduate work, I am an evidential medium who regularly channels departed loved ones for complete strangers. I bring through their gender, relationship, first names, personality, and a great deal of evidence that it is actually them, still alive, on the other side. All this is evidence that I simply could not have known, and for which the simplest explanation (Ockham's razor) is that I am merely communicating with the surviving consciousness of the deceased, who is now living in alternative dimensions.

Today (beginning in 2014), I also regularly participate in rigorous triple-blind scientific experiments on mediumship, working with Dr. Dean Radin, Ph.D. and Dr. Arnaud Delorme, Ph.D. at the Institute of Noetic Sciences, and with Dr. Gary Schwartz, Ph.D. at the Laboratory for Advances in Consciousness and Health (LACH) at the University of Arizona, Tucson. I will say more about all this in Section 5 of this chapter.

But for now, simply realize that as scientists we are proving that mediumship is very real. It is apparent through the research at both of these laboratories that mediums are receiving information from outside the body and brain, which the Ph.D.'s often monitor with thirty-two terminal EEG and computers that model what is happening in my brain while I am receiving veridical evidence. Mediumship is simply real and mediums can do it under rigorous, triple-blind laboratory conditions, as well as at circles, séances, and Spiritualist church services.

After one hundred years of psychical science, no skeptical scientist has produced any evidence whatsoever from any laboratory in the world that shows mediumship is not real. The only "evidence" any skeptic has is a childish wish not to believe it, just like the childish "wish" of fundamentalist Christians is that mediumship is "of the devil." When either of these "true believers" Christian Fundamentalist or *Materialist* pseudo-skeptic says, "All of this is untrue" or "This is of the devil," ask either of them if they have ever consulted a medium, and then watch them come unglued in almost identical fashion. It is then that you fully realize pseudoscience *Materialism* is truly a religion.

1. COMMUNICATION THROUGH MEDIUMS:

When we speak of mediums, we are talking about the oldest and most powerful form of psychical power. In the Middle Ages, this was called "witchcraft" by the *Materialist*-minded disbelievers, and is still thought to be "of the devil" by the Fundamentalist Christians. Like my sister, Joyce, who told me in February 2015 that as a medium, I need to repent or I "will burn in hell."

Unfortunately, my Fundamentalist sister does not realize that our beloved departed mother (who was also a Fundamentalist) has already contacted me through a medium (who was a complete stranger to me), and that our mom has actually asked forgiveness for her own rigid Fundamentalism. Now that Mom is on the other side, she explains that she realizes the afterlife is not at all how the church portrayed it to be. Indeed, it is not the "heaven" she was led to believe; it is just the next stage in a continuing saga called eternity. Yes, we are all eternal, but we are not stagnant; we advance only through our spiritual growth.

My mother has also apologized for leading Joyce astray. But she also says, *"Don't bother to straighten her out or argue with your sister. She'll find out soon enough when she gets here, and I will greet her. Part of my reparation for my rigid Fundamentalism is to welcome her to this side and then straighten her beliefs out through a lot of me eating crow."*

What is actually taking place in mediumship is not any form of witchcraft and has nothing to do with evil forces or imagined devils. The medium or "sensitive" is merely a person with a certain psychic talent that is often inherited and allows them to establish what physicists today would call a p-car resonance (quantum coherence) with the specific consciousness that is the subject of the desired communication. In other words, they are able to "dial up their wavelength." The Christian church is so against mediumship because it allows direct communication with Spirit, and the church needs to keep a meter on "talking to God."

One of the earliest cases of after-death communication (ADC) comes from the Bible's Old Testament in the Book of 1 Samuel: 28. This specific case bears close examination because of the conundrum it presents to Christian believers who, for doctrinal reasons, wish to discredit ADCs. This conundrum arises for Christians due to the following reasons:

- If every word in the Bible is literally true (as they want to claim it is), and
- If the Bible also clearly speaks of ADCs as being a reality (which it clearly does), then
- Communications with the dead must, by their own standards of biblical truth, be completely valid.

In this account, King Saul did not contact the devil, or Satan, He just had a medium call up the discarnate Samuel to get a prediction, something Samuel did when he was alive. Of course, most Fundamentalist Christians have never heard of this old passage in 1 Samuel 28.

But my point is that mediumship has existed in all cultures for all of recorded history. And in my own experience, it is completely valid. The following well-documented and famous cases from the twentieth century show how detailed it can be.

2. INTERESTING TWENTIETH-CENTURY AFTER-DEATH COMMUNICATIONS:

Here are several modern cases documented by ADC researchers, which you can look up online for yourself:

A. HMS BARNUM ADC:[48] During World War II, in 1944, an English medium, Mrs. Helen Duncan, was conducting one of her regular séances for a group of women in Birmingham, England. During one séance, she was spontaneously interrupted by the disincarnate consciousness of a recently drowned man who claimed to have been a sailor from the crew of HMS *Barnum* (a British warship). He contacted Mrs. Duncan at this séance because his mother was in attendance that same evening that he had drowned, and he wanted to report to her that his ship had just then been sunk by the Germans and that he was safely on the other side. Speaking through the medium, he told his mother that everyone aboard, including himself, had drowned.

Unfortunately, British Naval Intelligence did not want this classified information about the ship's sinking to get out, and so arrested Mrs. Duncan, accusing her of the crime of *witchcraft* under a law dating back over 200 years to 1735. Unfortunately, the Navy continued the prosecution and Mrs. Duncan was actually convicted in 1944 of this ancient crime of *witchcraft*, and was sent to prison.

Now, it is apparent from the facts of the case that Helen Duncan's only *crime* was that she somehow knew the truth and told the truth, which was being related to her by the disincarnate spirit of the drowned sailor. So, Helen Duncan was literally jailed for the "crime" of telling the truth.

British Prime Minister Sir Winston Churchill was adamant at the time that Mrs. Duncan had not committed any crime and should be immediately released. However, being that it was wartime, British Naval Intelligence won the day and kept Mrs. Duncan in prison until the war was over the following year.

Winston Churchill was so moved by this travesty of justice (jailing someone for telling the truth) that he then bothered to have all the "anti-witch" laws in Great Britain repealed in 1951, so that today in the United Kingdom, "witches" (mediums) have the same rights and freedoms as any religious clergy. And subsequently the Anglican Church has also accepted mediumship as being completely valid.

Finally, in 1998, fifty-four years later, according to a news story reported by Reuters, a full pardon was finally being processed for Mrs. Duncan. *To check this story online go to the HMS Barnum Association at www.hmsbarnum.com, click on THE SHIP, and then click on the HELEN DUNCAN STORY.*

The simple fact remains that:
1. Mrs. Duncan was actually communicating with a dead sailor and was receiving sensitive classified information from the other side in real time, and/or
2. British Naval Intelligence is a foolish organization that believes in "witches."

Frankly, I don't believe British Naval Intelligence is that foolish. But below is another example you can also check out online.

B. CHAFFIN'S WILL ADC:[49] The 1927 proceedings of the Society for Psychical Research document the case of a North Carolina farmer named Chaffin. Chaffin had written a will in 1905 that left all his money and property to his third son, and disinherited his other two sons and his wife.

Chaffin did not die until sometime after 1920, and the 1905 will was not his *last* will and testament. Unfortunately, after his death, no later wills were found and the 1905 will was the only one available. As a result, the estate was distributed to the third son.

Four years after Chaffin's death, one of the disinherited sons, James, began to have a series of vivid visions of his father during a night of restless sleep. The apparition of his father kept telling him, "*You will find my will in the overcoat pocket.*"

James was moved by these vivid dreams and later located his father's overcoat in the possession of the other disinherited brother. Together, the two brothers carefully examined the coat and located a paper sewn into the lining. Ripping out the lining they found a note card, which said, *"Read the twenty-seventh chapter of Genesis, in my daddy's old Bible."*

Their mother still had that Bible, which was now so old that it fell apart when they opened it. But it did contain a later holographic (written in his hand) will dated in 1919, which was clearly written in Chaffin's own handwriting. This will divided the property equally between the three sons. The handwriting of the will was checked and determined to be that of Chaffin. This more recent "found will" was not contested in court by the third son, who had inherited all under the 1905 will, but was allowed to stand with all three sons inheriting.

This case is remarkable because:
1. There was no living person who knew the location of that holographic will, or that it even existed.
2. It appears that it was the deceased father's consciousness, living in the afterlife, that wanted to set things right and contacted one of the disinherited sons four years later.
3. So, from beyond the grave, Chaffin took the effort necessary to communicate through a vivid dream to those still living.

To verify this story, Google: *Chaffin's Will.*

C. GEORGE PELLEW'S ADC:[50] A very famous trance medium in Boston, MA, Mrs. Lenora Piper, who was carefully studied by Dr. William James, Sir Oliver Lodge, and other members of both the British SPR and the American SPR, channeled the surviving consciousness of George Pellew, who had died previously. Mr. Pellew spoke with over 150 different people who attended various séances conducted by Mrs. Piper. Thirty of those people had actually known Pellew before he died. Mrs. Piper, however, had never met Pellew while he was living. In the physical, Pellew had always been skeptical of mediumship and the afterlife, but while living he had made a pact with one of the SPR investigators, Dr. Richard Hodgson, that "If I ever die and find out the afterlife is real and mediumship is valid, I will come through to you."

The discarnate Pellew, speaking through Mrs. Piper, was able to identify twenty-nine of the thirty people who knew him before he died, but none of whom were known to the medium. The only person Mr. Pellew could not properly identify was a childhood friend whom Pellew had not seen in decades. The people he identified easily and correctly were convinced by the mannerisms Pellew exhibited (through the medium) that they had in fact spoken with George Pellew, still living in the afterlife. Many were willing to testify that they had communicated with a disembodied, surviving consciousness that appeared to be George Pellew. This case is wonderfully told in Deborah Blum's 2006 book, *Ghost Hunters,* and also in Michael Tymn's 2013 book, *Resurrecting Lenora Piper.*

This brings us to one of the best-documented double-blind experiments, one that took place across the divide and was invented and carried out entirely by deceased SPR investigators from the other side.

D. FREDERIC W.H. MYERS ADC—The Cross Correspondences:[51,5] This is one of the most famous of all after-death communications (ADC) because it was orchestrated *posthumously* by one of the great minds of psychical research, Frederic W.H. Myers. Widely known as the *Cross Correspondences,* Myers devised this double-blind study which he actually conducted using ADC beginning six years *after* his own death. Myers sent scraps of messages back from the afterlife to five mediums on three continents, scraps purposefully delivered in foreign languages not familiar to the mediums. These scraps would only make sense if all were collected and pieced together by the British Society for Psychical Research (B-SPR). Here is that story:

Myers, who was a founding member of the B-SPR in 1882, died nineteen years later in 1901. A few years after his death (1907), Myers began sending messages back from the other side to the five mediums. Myers continued doing this for over thirty years—all posthumously. Eventually, two other members of the B-SPR, former colleagues of Myers, also died. Shortly thereafter, they also joined Myers in this project from the other side. The three of them actually collaborated in the afterlife and continued their work for the B-SPR together from beyond the grave.

These deceased B-SPR members sent parts of each message to the different mediums like pieces of a puzzle, quoting obscure Greek and Latin poetry that would be unknown/incomprehensible to the mediums, who did not know Greek or Latin. These mediums recorded the scraps through entranced (automatic) writing. Much of this writing made its way to the B-SPR in England, where it was subsequently pieced together.

> *To each automatist (a medium who does automatic writing) the information would be so fragmentary and strange as to be meaningless; but when pieced together, it could carry information of a kind that only could have come from Myers himself..." —*Brian Inglis [52]

This double-blind demonstration, designed and carried out by Myers, who was assisted by his deceased B-SPR colleagues, was purposefully transmitted so disjointedly in order to provide a double-blind communication that would silence all the skeptics simply by showing that it *could not* have been one medium's idea, but that it was actually Myers' consciousness communicating back from the afterlife.

Myers knew that if there was any way a skeptic could say that the mediums "imagined it," or that it was a hoax, or that they could read each other's minds, then the skeptics would attempt to do so.

To prevent this, Myers set it up where no single medium would know what the other was doing. What Myers and his colleagues did was no easy task: to orchestrate the activities of five mediums on three continents, and doing it all through the power of ESP. You can Google *Cross Correspondences* to read all about this.

5. RUNKI'S ADC:[53] In Reykjavik, Iceland, a séance circle lead by Hafsteinn Bjornsson had been meeting for séances in 1937 and 1938. In 1939, Runolfur Runolfsson, or Runki, a disembodied consciousness, began to come through at each meeting. Shortly before this consciousness came through, a new member named Ludvik had joined the circle. When Ludvik began to attend, Runki spoke through Bjornsson and revealed that in October 1887 (fifty-two years earlier), he had been out drinking with friends. On his way home that evening, he was drunk and so lay down and fell asleep on the rocky seashore. Unfortunately, the tide came in and he was drowned and then swept out to sea. Runki said, "I was carried in by the tide, but the dogs and ravens tore me to pieces." Later the remnants of his body were buried in a nearby grave, but his thighbone was missing. The bone, according to Runki, was "carried out to sea again, to later wash up on the beach at Sandgerti. There it was passed around, and it is now in Ludvik's house."

However, Ludvik knew nothing of any bone. Later inquiries among some of the oldest people in the community of Sandgerti remembered a very tall man's leg bone that had been found on the beach. Also, for reasons that no one could recollect fifty-two years later, the bone was reported to have been placed in the interior wall of the house now occupied by Ludvik. Subsequently, a large leg bone was retrieved from the inside wall of Ludvik's house. It was then verified that Runki had indeed been a very tall man.

Of course, now the question comes up: Why does the disembodied consciousness of Runki still need his thighbone some fifty years later? But the fact that this story illustrates is that Runki's consciousness was still continuing somewhere, half a century after his physical death.

You can Google Runolfur Runolfsson, or Runki to find numerous references to this incident. Also, you can go to http://www3. hi.is/~erlendur/english/mediums/gudni.pdf and download a PDF file of Erlender Haraldsson and Ian Stevenson's report on Runki published in the Journal of the American Society for Psychical Research, *Vol. 69, July 1975.*

E. MONTAGUE KEEN ADC:[54] Montague (Monty) Keen, age seventy-nine, one of Britain's more prominent researchers in psychic phenomena and a member of the SPR, died of a heart attack while speaking at the podium on January 15, 2004, during a public debate with a skeptic at the Royal Society for the Arts in London. A few weeks later, his wife Veronica contacted University of Arizona scientist Dr. Gary Schwartz Ph.D. and reported that she had received messages from her deceased husband through several different mediums, all stating that Monty wanted to conduct some research with Dr. Schwartz, whom he had met several years earlier. Dr. Schwartz and his research associate, Julie Beischel, then designed a two-part double-blind research project involving several mediums. One of the mediums they used was Allison DuBois, whose career as a psychic legal investigator is what the NBC program *Medium* is based on. During the session DuBois did not know anything about the subject of the reading, but Monty, speaking through DuBois, described his death as "falling at the podium," and then Monty referred to an upcoming event dedicated to his memory, an event that had been scheduled only after his death.

So, here we have the deceased spirit speaking about events that were not on any calendar before he died and were also unknown to the medium. Further, Monty said, speaking through the medium, that this event was a *"somewhat flattering surprise."* He also said that their communication in this way would be an excellent "white crow." This speaking about things that were not part of his memories before he died disproves the idea that mediums are merely receiving traces of a deceased consciousness memories from some Akashic record, but are, in fact, communicating with a living, surviving consciousness.

Dr. Schwartz then questioned the medium, who, it is apparent from the transcript, had no idea what significance the reference to "white crow" might have had. There were several additional facts conveyed through the medium that were unknown to both the medium and the sitters, but were known to Monty. This proves that it was his disembodied consciousness speaking from the other side through the medium in real time This entire séance was videotaped at the University of Arizona Research Laboratory.

Monty had been a member of the British Society of Psychical Research for over fifty-five years. I have read the published transcript of that DuBois reading. It is clear that Monty replicated the type of proof from the other side that Frederic W.H. Myers had provided in the famous Cross Correspondence work a hundred years earlier.

> *"...The thing about the afterlife that stood out for him, and that made him so happy is how he could still be here so much after his passing, and, how he would feel energy-wise, like he did when he was younger instead of with the issues he had accumulated as he got older."* [8] —Monty Keen, from the other side

The researchers concluded that the medium was definitely receiving information related to the designated deceased which was outside the possibility of telepathy. And they further stated:

> *"...These kinds of observations provide compelling evidence, if not convincing evidence, that intention, choice, and intelligence, and hence some sort of personal consciousness, survive bodily death."*[55]
>
> Gary E. Schwartz, Ph.D. & Julie Beischel, Ph.D.,
> University of Arizona 2005

Monty's wife, Veronica Keen, reports that there have been other communications from him recorded in England, New Zealand, Ireland, and the United States. He has reported that *his work is even more important* where he now is *than the work he was doing in the physical.*

3. *DREAM COMMUNICATIONS:*

THE BIBLE SPEAKS OF DREAM COMMUNICATIONS: In the Old Testament, the Bible speaks of Daniel interpreting dreams, as did the Old Testament Joseph (Yura) interpret dreams for the pharaoh when he was in Egypt. Jungian psychologists today provide the same service as Daniel in using the symbols of dream interpretation to study the subconscious mind. This is, in fact, very similar to using the symbols of the Tarot cards to interpret the desires of our hearts. Yet both Jungian dream interpretation and Tarot are passive methods of observation or "mind-reading" and only reveal our present perspective.

Further, in the New Testament, Joseph, the husband of Mary, the mother of Jesus, twice got messages through dreams. So it is evident that the Bible endorses communications from spirits through dreams as being a valid medium.

DREAM COMMUNICATION IS OPEN TO EVERYONE: Each of us is gifted with this psychic ability to dream, and our dream symbols can also be manipulated by a deceased consciousness to communicate with us. So, you might ask, "If dream communication is so simple, then why haven't my own dead relatives communicated with me?" The short answer is that you have not been making yourself accessible to them.

When the consciousness of the living does not make itself accessible, it is more difficult for the consciousness of a departed loved one to communicate. For example, it took Chaffin four years to finally get through to his son through a dream about the missing will. For four years, he had a very important message to get across, but he had to learn how to do this from the other side.

You can contact your deceased relatives in this same way. It just takes a strong belief that they can hear you and will also contact you, and being willing to face whatever comes—with an open heart.

The Hasidic Jews have been studying this phenomenon and working on this doctrine for several hundred years and have formulated some fairly firm conclusions. Rabbi Yonassan Gershom said in his 1992 book[56] that there do appear to be some common factors that hold spirits earthbound; these factors, as described by Rabbi Gershom, fall into three main categories:

1. **The spirit does not know it is dead:** While it may seem incredible that someone could die and not realize it, this does happen. Such a soul often has no belief in life-after-death, and therefore reasons that, because it still has consciousness, it cannot really be dead. Also, if the death happens suddenly, the transition to the afterlife can be so fast that the soul does not realize it has taken place.

Note: My own personal NDE (1970) left me with memories of waking up with the Light, but never passing through a tunnel or leaving the body. I definitely remember returning into the body, but I have no memory of my actual death. Consequently, I can understand that a surviving consciousness could be dead and not realize it.

2. **The soul has stereotyped expectations of the next world:** Many people simply do not realize that their religious descriptions of the afterlife are only metaphors or speculations. When they die, they expect their beliefs to be fulfilled *literally*. For example, the absence of a heaven with angels plucking harps may cause some Christian souls to remain earthbound because they will not accept the help of the spirits who have come to meet them. And in some cases, they may believe that the helping spirits are demons in disguise, because that is what they were taught to expect by their overly confident preachers who are so sure their ancient speculations are correct.

Also, when an atheist person dies who believes that **when you are dead you are dead,** they realize they are obviously still alive (although no one seems able to see or hear them). But believing that you cease to exist after death, and since they still exist, they surmise that they must not then be dead.

Edward C. Randall (1906) assisted numerous earthbound souls to move on during his years working with Mrs. French. He says that half of every evening's sessions for over fifteen years were devoted entirely to this "mission work," assisting earthbound spirits to move on toward the light. In several cases, there were Christians literally, as Rabbi Gershom said, waiting for the Savior to show up, and believing that the spirits of their dead relatives were imposters.[57]

3. **The soul has unfinished business:** This is the most difficult type of earthbound spirit to work with because in order for a soul to be released into the light, it must want to go there. As long as it feels that it has not completed its work here on earth, the soul will continue to be attached to persons, places, or things in the material world.

Note: This is why most near-death survivors return to their physical bodies; they have unfinished business here in the physical which they return to complete before finally moving on.

4. CENTURY-OLD AFTER-DEATH COMMUNICATION VERIFIES MODERN NDE:

THE FRENCH REVELATION: In 1906, Edward C. Randall, a well-known New York lawyer, published his findings after fifteen years of research into after-death communications (ADC) working with one medium, Emily S. French, using the ADC phenomenon called direct voice. Here is what he received from a deceased individual's surviving consciousness communicating to him through the mediumship of Mrs. French, nearly *a century before the near-death experience was recognized as valid* by the medical community.

Randall's reports have been researched and re-compiled by N. Riley Heagerty in a book entitled, The French Revelation.

DR. DAVID C. HOSSACK'S[58] **DEATH PROCESS as reported in 1906: This description clearly matches the events of the modern near-death experience, yet this description came through during séances in the years before 1906.** Randall had spent many years talking to disincarnate spirits from the other side while working in the presence of medium Mrs. French. But living in the late nineteenth and early twentieth century, Randall had no opportunity to have heard of the modern idea of the near-death experience. And at that time, the ancient accounts of the NDE were obscure ancient literature. Yet what Randall reports as being received from a surviving consciousness is *identical* to the NDE.

The following is a published description of the death process and awakening into the afterlife, as reported to Edward C. Randall by Dr. David C. Hossack, who had died seventy-one years earlier in 1835.
(Parentheses are the author's comments.)

> *"There was this meeting and greeting with my own (family) who came to welcome me, as naturally as one returning after a long journey in the earth-life would be welcomed. Their bodies were not so dense as when they were inhabitants of earth, but they were like my own. Then I was told that my body and the bodies of all those in that life were actually the identical bodies that we had in earth life divested of the flesh covering. I was also told that the condition was a necessary precedent to entering the higher life, and that such bodies in earth life had a continuity and, further, that in leaving the old, I had come into a plane where all was etheric* (dark energy), *that is matter vibrating in perfect accord with my spirit, technically speaking, the "etheric self"* (dark matter spirit body). *To me everything seemed perfectly natural to sense, sight, and touch.*

110

"Again, let me tell you that the outer flesh garment is not sufficiently sensitive to feel, the etheric body (spiritual body) *alone has sensation... I found little body change. I had sensations and vision and my personal appearance was in no way changed except that my body was less dense, more transparent as it were, but the outline of my form was definite, my mind clear, the appearance of age gone, and I stood a man in the fullness of my mentality, nothing gained or lost mentally.*

"What impressed me most after meeting with my own (family and friends) *was the reality and tangibility of everything and everyone. All those with whom I came in contact had bodies like my own, and I recognized friends and acquaintances readily. Now, I will tell you of the one thing that impressed me most on coming here. That was that matter in its intense refinement, in its higher vibration (as observed here), was capable of intelligent thinking and direction. Shape and grasp this proposition if you can; I could not in the beginning, nor could I comprehend at once that* **all in the Universe was life and nothing else.***"*

Note: *Here Hossack is stating a very Buddhist principle that earth, plants, insects, bacteria, animals, and humans all are the same life. Think "Leaves of Grass" by Walt Whitman.*

"This fact, which we now know (in the afterlife), *will overturn the propositions of science* (in the physical life i.e. 3-D space-time).

"In all the orthodox teaching of nearly two thousand years, not one law has been given tending to show how it was possible for individual life to hold continuity. Theology had claimed it without explaining how or where. This no longer satisfies the human heart or mind, a fact which accounts for the great unrest amongst your people in every land. For this reason it has been our aim to explain the law through which life is continued, and so simply to state the facts and explain the conditions that all may understand.

"The key to comprehension is first to realize that your Earth *(reality as you perceive it)* **DOES NOT CONTAIN ALL THE MATTER OF THE UNIVERSE, that all that you see and touch is but the substance used by life in growth** (i.e., light energy).*

"When one leaves the earth-condition (light energy), *divests himself of the physical housing, he, through such change* (into dark energy), *ceases to be mortal. By becoming a resident of the new sphere* (dark energy) *he is said to take on immortality, but in reality,* **he has always been immortal.**

"You regard the telephone as wonderful, wireless telegraphy (radio) *more wonderful still, but we communicate with each other by simple thought projection* (ESP). *You regard the phonograph as a marvelous instrument, but it is crude beside the instruments in use among us. When you appreciate the truth that* **we live in a state no less material that your own** (dark matter), *you will understand that with our greater age and experience we are much in advance of you, and make and use appliances and instruments that could hardly be explained to mortal mind* (which cannot yet even discern dark energy or dark matter). *At another time I may be permitted to discuss the subject more fully."* [59]

Note: *In the bold text here, Dr. Hossack is talking about the quantum mechanics of string theory and alternative dimensions, which did not begin to enter modern physics until thirty years after Randall published this account of Hossack's revelations.*

SCIENTIFIC REPLICATION: For me, the fact that Randall, who, at the time, could know nothing about the near-death experience, was reporting this same phenomenon in 1906, some seventy years before modern studies of near-death experiences began, provided for me effective replication through double-blind conditions for purposes of comparison with the modern NDE.

A century earlier, Randall was delivering the same information that now occurs in all modern NDEs. This is compelling evidence of two things:
- First, it is replication of my own experience; and
- Second, this replication itself is virtual proof that Randall was actually communicating with a surviving consciousness on the other side, one who had also been through the same death process as described by the NDE.

The only difference between what Dr. Hossack experienced and what near-death survivors are reporting is that Dr. Hossack *did not come back to his prior life,* **as all the NDE people have done.** Instead, Dr. Hossack is reporting the same after-death experience while he is now a surviving consciousness in the afterlife. This amounts to scientific replication of my own experience that appears to prove two things:
1. The NDE is, in fact, the same process that is actually experienced at death.
2. Mediums do, in fact, communicate with disincarnate spirits (surviving consciousnesses) after death.

The simple fact that a disincarnate surviving consciousness knew things about the death experience, and also about quantum electrodynamics and the conscious universe that Randall could not possibly have known in 1906 is strong evidence of the survival of Dr. Hossack's consciousness in a separate reality.

This disincarnate consciousness communicating with Randall purported to be the surviving consciousness of the late Dr. David C. Hossack. My own historical research shows that Dr. Hossack was a former professor at Columbia University (King's College), and founder of the Columbia Medical School. Dr. Hossack's interest in physics is in keeping with his position as a professor in his former physical lifetime. And so it appears that Hossack got together with other like-minded consciousness survivors, on the other side, and studied these things there. Much of what he related involved quantum mathematics and quantum electrodynamics, which was not known when Dr. Hossack himself was alive, and also was not known when Randall wrote about the conversation in 1906, but would be discovered a quarter-century later.

> **Note:** My further research on Dr. Hassock has found that he graduated from Princeton University and received his medical degree in Philadelphia in 1791, and subsequently studied abroad at Edinburgh, Scotland, before returning to New York. Dr. Hossack was one of the founders of the New York Historical Society in 1804, and also established the Elgin Botanical Garden between Forty-seventh and Fifty-first Streets and between Fifth and Sixth (Avenue of the Americas) in Manhattan.

WHAT IS THE AFTERLIFE LIKE: Randall went further and asked several disincarnate spirits about what the afterlife is like:

"The so-called dead live here about us... The substance that forms the bodies of spirit-people, vibrating at more than five octaves higher than the violet-ray (violet = 7.5 x 10^{14}hz, double that five times), *few in earth life ever see, though spirit-people see and talk with each other and with mortals when the necessary conditions are secured... I know that every hope, ambition, and desire of earth are continued beyond this life, as is also the burden of wrong (karma). I know that we are as much a spirit now as we ever shall be, that in death, so called, we simply vacate and discard this gross material that gives us expression in the physical plane."* —Edward C. Randall [60]

Randall's description of beings in the afterlife, living right around us but at a higher vibration, coincides with quantum field theory (string theory), which postulates eight additional hidden energy fields (M-theory) that exist "around us" but have not yet been discerned.

Randall's reports of what the people on the other side told him are strikingly similar to what the "Masters" are reported to have told Dr. Brian Weiss nearly a hundred years later. Randall's reports also agree with what the near-death people have been reporting in the twenty-first century about how the "veil is lifted" and reality is glimpsed. In each of these separate cases, when the consciousness is leaving the physical dimension and going to an alternative reality, they all report hearing a noise that sounds like a rushing wind.

As a student of physics, I speculate that this rushing wind sound is caused by the speeding up of their vibrations to suit the higher frequencies of the new plane of reality to which they are going, for example:

1. Thoth-Hermes in 2500 BC described these alternative realities as seven planes with many levels in each plane. He also said everything was made of vibrations (string theory).

2. The "Masters" told Dr. Weiss that there are seven planes with many levels.

3. A surviving consciousness from the other side told Edward C. Randall that:

> "There are innumerable spheres in the spirit world; if it were not so progression would be a myth... Some tell you that there are only seven. That is because they have no knowledge beyond that sphere (seven). I do not mean a place fixed by boundaries, for the spheres or degrees in spirit life are only conditions and are not confined to a limited space... As a soul develops, it naturally arises above its surroundings, and consequently experiences a change in its spheres or conditions (chakras)."

DARK ENERGY: Just what are these alternate dimensions where these surviving consciousnesses exist and apparently surround us? The quantum electrodynamics of the late twentieth century points to the fact that the universe is not made of empty space, as we like to imagine it, but that instead all of space is filled with dark energy. An energy force that we can prove by using astronomical mathematics must be there, but which we have not yet been able to discern. What this energy might be we cannot yet measure with our crude instruments. Yet back in 1906, before Einstein's theory was published, Edward C. Randall had this to say after he held ADC discussions with Michael Faraday and Dr. Hossack:

> *"They believe the spiritual plane is filled with ether* (dark matter) *similar to earth substance* (light matter) *but in a very high state of vibration... According to them, the universe is all material substance or matter in different and varying states of vibration."*[61]

Indeed, this is a statement of the alternative dimensions postulated by string theory. Further, Dr. Hossack,[62] in another after-death communication, told Randall that:

"The most learned scientist among the inhabitants of earth has practically no conception of the properties of matter, the substance of the universe, the visible and the invisible. I did not when I lived among you, though I made a special study of the subject. That which you see and touch making up the physical or tangible, and having three dimensions (3D), is the lowest or crudest expression of life force and not withstanding my long study of the subject, the idea that the physical (matter) had a permanent life form or that what you call space was composed of matter filled with intelligent and comprehensive life (conscious universe) *in a higher vibration* (separate energy field) *never occurred to me. So when I became an inhabitant of the plane where I now reside, I was wholly unprepared to grasp or comprehend the material conditions of the environment in which I found myself."*[63]

The subject under discussion is clearly the zero-point energy (ZPF) field and dark energy, which would not be discovered until many decades later. Now, whether Randall was actually speaking with Drs. Hossack and Faraday is up to you to decide. But Randall, a New York lawyer and not a physicist, is obviously reporting something completely unknown to nineteenth-century materialist science and that we only began to discover nearly half a century later, but it is something those same gentlemen claiming to be speaking to him would have understood completely when they were living in a fourth-dimension energy field called the afterlife. This brings us now to the subject of quantum energy fields, which will be discussed in the following chapter, but first let's look at my own experience of mediumship.

5. MY PERSONAL RESEARCH AND EXPERIENCE OF MEDIUMSHIP:

As a scientist, I finally decided to investigate mediumship from the inside by becoming an evidential medium myself. So, in 2006, I enrolled in the **Morris Pratt Institute** course on Modern Spiritualism, which covers the philosophy of Spiritualism, mediumship, and spiritual healing. I completed the four years of graduate-level course work in 2013. I then joined the *Licentiate Ministers and Certified Mediums Society* of the National Association of Spiritualist Churches and began to serve regularly as the medium providing "Spirit Greetings" (three- to five-minute readings) at church services for two local San Francisco Spiritualist churches.

I have also traveled twice to **Arthur Findlay College of Psychical Science,** outside London, to study mediumship. And I continue to spend two weeks there every year. This is one of the finest places to study mediumship in the world. It is a Victorian estate that belonged to Arthur Findlay and is devoted entirely to the advancement of mediumship. It is the graduate school for mediums. If you are familiar with the Harry Potter novels, this is *Hogwarts Academy,* and I will speak more about it later.

Here is the URL for a ten-minute video about the college: https://vimeo.com/104742981

But during my first time on the platform as a medium at an American Spiritualist church, I immediately began to bring through evidential information that I could not have possibly known, except that it was being given to me from Spirits on the other side. This is evidential information that is verified by the receiver (sitter).

One "greeting" during that very first time serving as a medium was from a deceased former boyfriend of a young woman in the audience. I gave her these facts:

1. There is a young male figure, approximately your own age, who knew you in school.
2. He has black curly hair,
3. He liked to grow his beard and then shave it off, as young men do, only to grow it again.
4. He is not your relative, but a dear friend.
5. He had something to do with basketball.
6. His message is that he loved you and wishes he could have spent more time with you in this life.

Afterward she came up to the platform and said, *"I know exactly who that is, but he had nothing to do with sports or basketball. His name was Jacques. He is a former boyfriend and is deceased. So his message is very relevant."*

The following morning, I asked my chief guide, Helen, what the basketball was about. She said to me from spirit, *"Oh, Alan, the only professional basketball player you ever knew personally was Jack Sigma, of the Seattle Supersonics, and I was just trying to give you his name as Jack (Jacques)."* So, now we have agreed to use this symbol. If I see basketballs, I will say, *"Did this spirit have something to do with basketballs, or was his name Jack?"*

But later, in March, of the following year, after attending Arthur Findlay College, and as my ability to release and let spirit speak through me improved, I was able to bring through this same woman's grandmother by name.

MEDIUMS HAVE GUIDES: One of the most fun parts of being a medium is the fact that we develop relationships "beyond the veil." Now, I "talk" every day with my mother-in-law, Helen, who has been deceased for twenty-three years as of last December. The truth is that I actually never met Helen in this physical lifetime, as she made her transition five years before I met and married her daughter.

But Helen has come to me of her own volition, through many different mediums on different continents, and it was she who initiated communication with me from the other side. I never went seeking her; instead she has identified herself as my designated "chief guide." And invariably whenever I consult a medium, from any country, they immediately bring through Helen with undeniable evidence they could not have known, including her exact name, what she looked like, and where she was born.

She is also very persistent in our relationship and wakes me up at 3:00 every morning so I can meditate and she can speak to me through inspirational writing (not automatic writing). This has continued now for three years.

I have also heard from many of my deceased relatives through various mediums who do not know me from Adam, but who have brought through my father, my mother, my uncle Bob (by name), my grandmother Lucy, who has been deceased for fifty years, and my grandfather (Lucy's husband) "Papa," as well as non-relatives who were my friends. Always these communications are accompanied with undeniable evidence that the medium delivering the message could not possibly have known. All of these people are working as my guides, as well as numerous unidentified discarnate scientists.

MEDIUMSHIP IS A LIFELONG ENDEAVOR: Although I did not see it going in, mediumship is not a weekend experiment; it is a lifelong work. Like a student concert pianist, I will continue my study and practice of evidential mediumship by annually taking "graduate" courses at Arthur Findlay College of Psychic Science, outside London, England. And completely at my own expense. I don't charge for my mediumship (but, please don't think I give free psychic readings to the public), and actually spend thousands of my own retirement funds annually perfecting my skills. Consequently, only an abject fool would make the charge that "all mediums are charlatans."

Also, to further the science of after-death communications (ADC), I currently participate as one of the research medium subjects at the Institute of Noetic Sciences, in Petaluma, CA, (noetic.org) in their ongoing scientific research into mediumship, as discussed below; and at the Laboratory for Advancement of Consciousness and Health at the University of Arizona, Tucson; and at the Psychic Research Foundation, which is mostly online now.

Frankly, I believe that all people have this mediumistic ability to receive messages from the other side, simply lying dormant and unused. Once we learn how to turn off our five senses, we can relate to our sixth sense. I know from my own experience that this was not a "gift," but entirely a learned skill that I acquired. I may have been motivated by my near-death experience, but I had to do the work of learning how to do it myself. Which is why I also today host a student-mediums circle, to help others to learn this same skill. The only "gift" seems to be your ability to believe that you have this talent.

ENERGY HARMONY: One of the most interesting facets of mediumship is the need for harmony. All the participants in a circle (séance) need to be on the same wavelength. Many skeptics who have attended mediumship circles and séances at Spiritualist church services, or in a private reading, have not received evidential messages and therefore believe that none of this can possibly be true.

But it is actually their own skepticism that creates the negative results. This seeming phenomenon is completely in accord with QED's observer principle, which says, *"What you see is what you get."*

These poor skeptics come to mediumship with a "show me" attitude, and so they go away empty-handed, with the same attitude, saying, *"I knew it wasn't true,"* never realizing they create their own self-fulfilling prophecy. Intent is so important in this work. If they could only believe first, then they could show it to themselves, but because they mentally choose not to believe, no one can overcome their skepticism and prove it to them.

Good communication with deceased spirits actually involves four individual consciousnesses. First, there is the deceased loved one who is communicating to a second spirit, who is the deceased medium (in my case Helen Bradley). Then, the deceased medium (Helen) is communicating with the medium (Alan) who is living in the physical. Helen does this by impressing thoughts on my heart when I am demonstrating as the medium here in the physical life. She sends them to me as a feeling because I am clairsentient (I am very sensitive and feel things). Next, I interpret those feelings into words, which I speak and are finally heard by the consciousness of the receiver (sitter).

But if any one of these four participants is out of sync, then the message will be lost. It is like four people at a party playing the game *Whisper Down the Lane* or *Telephone,* each whispering a secret message in the next person's ear, with the fourth person finally repeating it out loud. This is always fun because it comes out so garbled at the other end that it makes us all laugh. But if any one of the four refuses to play the game, then the entire message will not be transmitted. Mediumship works exactly like that, and when the receiver (sitter) has a "show me" attitude, there will be no message received.

Even when everyone is in sync, it can still come out slightly garbled, because I as the medium may try to analyze and make sense of it and so misinterpret it. Instead, I strive mightily to just say what I get with no coloration and no interpretation, allowing the receiver (sitter) to interpret the evidence I receive.

I count myself as entirely lucky to have had the near-death experience, which convinced me that the afterlife was real, before I ever began to try to communicate with the departed. My own NDE made it easy for me to believe that all of them are still alive after death, just like I had also been alive on the other side during my NDE. Consequently, talking to them was just the next logical step for me.

Now, I will share two examples from personal experience as a medium of what happens when all four players (**you** the sitter, **me** the medium here in the physical, **Helen** the medium in the afterlife, and **your departed loved one**) work together.

Here is what I got in a reading on **September 10, 2014** (names changed for privacy).

ALAN: *Bill, I have a man with a 1930s snap billed hat and curly brown hair… He says his name begins with a J… He died of lung cancer at age sixty-three… He is not a relative… Does that make sense to you?*

BILL: *Yes, definitely…I know exactly who he is.*

ALAN: *His message is simply that he still loves you and to just hang in there, buddy; it will be okay.*

BILL: *Alan, I couldn't remember his name when you said "J," but now it came to me… His name was Joseph… He was a fellow cab driver with me and we called those hats "cabby hats." He also had curly brown/reddish hair, which he hid under that hat; and his signature statement was "My hair is brown not red," which he would often say as he pulled his hat down to hide his hair. And he also died of lung cancer about that long ago.*

So we have these verified facts:
1. His name started with a "J."
2. He is not a relative.
3. He had brown hair.
4. His hair was curly.
5. He wore a snap billed hat.
6. He died of lung cancer.
7. WOW EVIDENCE: He always said, "*My hair is brown not red.*"

Of course, these are answers to the exact questions that I ask of each spirit before I will accept their message. There are fifteen questions and they usually answer the first six to eight until it is obvious that it is them.

1. What was your first initial? or What was your first name?
2. What did you die of?
3. What generation do you come from?
4. What was your relationship to the sitter?
5. What color was your hair? What were your personal characteristics (i.e., hairstyle, glasses, cane, etc.)?
6. Give me some wow evidence…an important fact that identifies you.

Then again on **October 15, 2014,** I brought through Grandma Mary "Mame" Adelaide, for sitter Mary Franklin (who has given me permission to use the real names).

ALAN: *Mary, although I know you well, I still know nothing about your family*
or who you have on the other side. I just know that you are married to Michael.

MARY: *Yes.*

ALAN: *Mary, I have a woman coming through from your grandmother's generation, on your father's side, maybe your great-grandmother's generation. She has an "M" for a name, and I want to say her name is Mary but then because that is your name, I am confused. Maybe she is instead Mame or Margo or Marguerite; she seems to be telling me that this is her other, or middle name, which she is now giving me… I feel that she died of a pain in the chest, because she is making my chest hurt on the surface here…maybe breast cancer… She had brown hair which she wore pulled back into a severe bun… Pulled back behind her head… She apparently died in the 1890s, because she is giving me that number: 1890.*

Her message is that she loves you…even though she never knew you… But she is happy that you returned to Spiritualism.

Does any of that make sense to you?

MARY: *Yes, Alan. That makes a lot of sense to me… My grandmother on my father's side's name does start with a "M" and it is in fact Mary, as you first thought. Indeed, I am named after her… I believe that she died of breast cancer, but I'll check it out… Actually, you probably thought great-grandmother because of the age difference. You see, I was born quite late in my parents' lifetime, and she died before I was born… So just like she said, I never knew her.*

1894 PHOTO OF GRANDMA MARY (MAME) ADELAIDE

Later, relatives verified that Grandma Mary's nickname (or, as I said, "other name") was "Mame." Officially she died of pleurisy, which is an inflammation of the chest cavity. These relatives sent her picture and verification of the following facts:

1. SPIRIT'S ACTUAL NAME = Mary (odds of better than 200 to 1)
2. *HER NICKNAME = Mame (1 in 25?)*
3. *GENERATION = grandma (1 in 10)*
4. *RELATIONSHIP = father's side (1 in 2)*
5. *WHAT SHE DIED OF = inflammation of the chest cavity (1 in 50?)*

6. *HER HAIR COLOR:* = *Brown (1 in 5?)*
7. *HER HAIR STYLE* = *pulled back into a severe bun (1 in 20?)*
8. *WOW EVIDENCE* = *She never knew sitter (dies before sitter was born). (1 in 25?)*
9. *WOW EVIDENCE* = *She is showing me 1890; picture was taken in 1894. (1 in 25?)*

Again these evidential facts are simply the answers to the exact questions I automatically ask of every spirit and demand answers for before I will accept the greeting or message. Statistically, the odds are 200 x 25 x 10 x 2 x 50 x 5 x 20 x 25 x 25 = **312.5 billion to 1.**

So, how does this happen? Obviously, I could not guess with that kind of accuracy. But if (as many people wish to claim) I am reading the mind of the sitter and cherry-picking these facts from their mind, just how am I doing that? If such amazingly accurate mind reading were possible, then all ESP must also be a fact as well. But both mind reading and guessing are too fantastic to be believed as answers to how this happens.

Also our scientific study of mediumship—at IONS Consciousness Research Lab, and at the Laboratory for Advances in Consciousness and Health (LACH) at the University of Arizona—has shown the information is coming from outside the medium's brain, while the medium is in an altered state but not actively thinking and is instead just receiving.

Clearly, the most logical explanation is simply that these facts are being provided by surviving consciousnesses who are actually speaking to me. Especially when I have given Spirit that precise list of fifteen questions which I require the discarnate loved one to answer, and they are obviously giving me those same specific answers.

A QUICK LOOK AT THE NUMBERS: On Sunday, March 22, 2015, demonstrating mediumship at the morning service at Celebrate Life Church in San Francisco, Spirit came through as disincarnate spirits for three separate sitters. Each time Spirit gave me correct first names, genders, and relationships. After the service I was awed to have achieved a statistical 100 percent.

Then again on Wednesday, April 29, 2015, when I was demonstrating mediumship at an evening service at Golden Gate Spiritualist Church in San Francisco, Spirit correctly delivered through me five first names of the five disincarnate spirits, all of whom were quickly placed with sitters, again 100 percent.

To guess these five names (Frank, Martha, Elizabeth, Florence, and Carla) correctly, while also providing gender and relationship, works out to about

200 to 1 for each guess. Yet most observers of this phenomenon don't do the math. To be guessing at odds of 200 to 1, and to then do it correctly five times in succession is simply impossible (200 x 200 x 200 x 200 x 200 = 3.2e+11 (320,000,000,000) or odds of 320 billion to one.

It is obviously not me doing any of this. I could not possibly guess that well. Instead, it is another consciousness, or in this case several surviving consciousnesses working through me. So, I am just the telephone.

SONNY COMES THROUGH: Here is another example that shows it is not ESP between minds in this physical life. While at Arthur Findlay College in December 2014, my tutor had me up on the platform in front of thirty-five people and wanted to "stretch" my abilities. She (Janet Marshall) said to me, *"Alan, do you know anyone that is in spirit, who identified with a cartoon character?"*

I answered, *"Yes."*

My tutor then said (with the command voice of a Marine Corps drill sergeant), *"Okay, link with them right now and bring us a message from them."*

For me this was an "Oh my God" situation. I remembered that Sonny Gee, who had died the previous year, identified with Mickey Mouse, but I had never met Sonny in this physical life. I just knew his widow, Carla Gee. I also knew that Sonny had been an evidential medium and an alumnus of Arthur Findlay College, and Carla had told me of their joint identification with Mickey and Minnie Mouse. But I had never brought Sonny through from Spirit.

Yet the tutor had just given me a "direct order" to link with a specific spirit.

So, I immediately shifted into the power, and I linked with Sonny. He now delivered a sermonette for the thirty-five in the audience, through me. It was quite emotional for me, as Sonny proceeded to deliver an eloquent mini-sermon through my channeling of his thoughts. I could not have conjured up this sermonette under those "do it now" conditions.

Speaking through me to the entire audience, Sonny now said,

> *"How many of you student mediums have been to see the Mickey Mouse characters who greet visitors to Disneyland and Epcot Center?"*

And although we were in England, with an audience of Europeans, more than half that class, from all over Europe and Australia, raised their hands, and then Sonny gave me this:

"Those Mickey Mouse greeters do not get emotionally involved when they greet and hug each little child coming through the gate... No, they are thinking 'When is my next coffee break?' But each of them, without getting emotionally involved, is actually delivering the message of love to each child... And that message is clearly coming from the departed soul of Walt Disney himself... Each actor then is unknowingly a direct medium for Walt Disney... Each of you as mediums also must not get emotionally involved, but can do your job while not taking on the emotions, just as well as those Mickey Mouse greeters do, while you too are thinking about your coffee break."

The message was a great object lesson for the class and also for me, because I tend to become overemotional when I work for Spirit. So, I know that it did not come from me, but came entirely from Sonny Gee, alive on the other side. It was a communication for the class to instruct them in how to *not take in the emotions* but still get the job done. This subject had come up in the class discussion in the prior hour, which also showed that Sonny was alive and listening in to the discussion.

So, my tutor's command, *"**Link to Spirit and bring that person through**,"* set the stage and allowed Sonny to preach once again from eternity. Simultaneously, for me, was the opportunity to demonstrate how to link with Spirit. Frankly, I knew Sonny's widow Carla, but I had never met him in this physical life.

Looking at this scientifically, this was not a sermonette that I could merely think up off-the-cuff, and it was also not something I had ever heard before. I could not have been reading anyone's mind because no one there except me had ever heard of Sonny Gee, and I had no memory of what he might have been like or what he might have said. I only knew that Carla and he often thought of themselves as Minnie and Mickey Mouse. Since it did not come from me or anyone else in the room, it had to have come directly from Sonny.

Honestly, I am not always able to achieve 100 percent, but overall I do well above 75 percent with names and relationships just by ignoring my analytical mind and relying on my sixth sense, which allows Spirit to demonstrate its existence through me.

6. SCIENCE INVESTIGATES AFTER-DEATH COMMUNICATIONS: 1883–1916, INVESTIGATIONS BY SIR OLIVER LODGE FOR SPR:

Sir Oliver Lodge—physicist and one of Great Britain's foremost scientists, who also holds patents in the development of wireless telegraphy (radio)— was elected as a Fellow of the Royal Society in 1887, and knighted in 1902. Lodge began investigating mediumship in 1883–1884 for the Society of Psychical Research (SPR) and continued this investigation until his death in 1940. He scientifically investigated all the great mediums, including Eusapia Palladino, Lenora Piper, and Gladys Osborne Leonard. He wrote about it in his 1909 book entitled, *The Survival of Man,* saying:

"The first thing we learn is continuity (i.e., life continues). There is no such break in the conditions of existence as may have been anticipated; and no break at all in the continuous and conscious identity of genuine character and personality. Essential belongings, such as memory, culture, education, habits, character, and affection. All these, and to a certain extent, tastes and interests, are retained.

"Meanwhile, it would appear that knowledge is not suddenly advanced, we are not suddenly flooded with new information, nor do we at all change our identity; but powers and faculties are enlarged, and the scope of our outlook on the universe may be widened and deepened. If effort here has rendered the acquisition of such extra insight legitimate and possible."

His life's work and his investigation of mediumship make great reading, because six years after Sir Oliver Lodge published this 1909 book confirming that science had already proven consciousness survival, his son, Second Lieutenant Raymond Lodge, was killed in action in Flanders, on September 14, 1915, giving Lodge the opportunity to personally verify the reality of consciousness survival.

Eleven days after making his transition, on September 25, Raymond Lodge began communicating with Sir Oliver and Lady Lodge through the mediumship of Gladys Osborne Leonard and Alfred Lord Peters.

By the end of April, 1916, a preponderance of evidence that Raymond had been communicating with them had been accumulated by the Lodge family. Sir Oliver Lodge then published a second book, Raymond or Life and Death, in which he said:

"The number of more or less convincing proofs which we have obtained is by this time very great… I am as convinced of continued existence on the other side of death as I am of existence here… It may be said, you cannot be as sure as you are of sensory experience. I say I can. A physicist is never limited to direct sensory impressions; he has to deal with a multitude of conceptions and things for which he has no physical organ—the dynamical theory of heat, for instance, and of gases, the theories of electricity, of magnetism, of chemical affinity, of cohesion, aye, and his apprehension of the ether itself, lead him into regions where sight and hearing and touch are impotent as direct witnesses, where they are no longer efficient guides."

2012–PRESENT, INVESTIGATIONS BY INSTITUTE OF NOETIC SCIENCES (IONS):

The recent IONS study published in 2013 investigated the correlations between the accuracy of mediums' statements and their brain electrical activity, and also differences in mediums' brain activity between four subjective states were studied, including: PERCEPTION, RECOLLECTION, FABRICATION, and MEDIUMSHIP communication. It was found that the *mediumship communication* mental state differed from the *perceptive* mental state, with larger amplitude high gamma power observed during the *mediumship communication* mental state.

This rise in gamma waves seems to be from eye, ear, or muscular activity. In other words, the mediums were receiving information through eyes, ears, and feelings, rather than imagining or fabricating them. The report concluded:

> *"The study's findings suggest that the experience of communicating with the deceased may be a distinct mental state that is not consistent with brain activity during ordinary thinking or imagination."*

This is important scientific verification of the fact that as *mediums we are actually* receiving *outside information* during *medium communication* mental activities. The mediums all propose that they are communications being received from the deceased. Although the data is insufficient to confirm scientifically that the source of the communications is actually surviving consciousness, it does show that the source is outside the medium.

The preceding are the graphic printouts from the 2012 IONS report showing the thirty-two-point EEG measurements of what mediums are doing in PERCEPTION, RECOLLECTION, FABRICATION, and MEDIUMSHIP, and as you can see MEDIUMSHIP is different from all the others.

The next step is to figure out experiments that can isolate data under test conditions that will explain where that information being received by the mediums is actually coming from. We mediums may "know" that it is coming directly from deceased loved ones, but science asks the question, **How do we prove that it did not happen some other way; like ESP from living minds?**

I was personally involved in a follow-on study, serving as a test medium at IONS in November 2014 and February 2015, where I worked with Dr. Dean Radin and Dr. Arnaud Delorme to verify what mediums are capable of, and I am currently involved as a sponsor of this ongoing work.

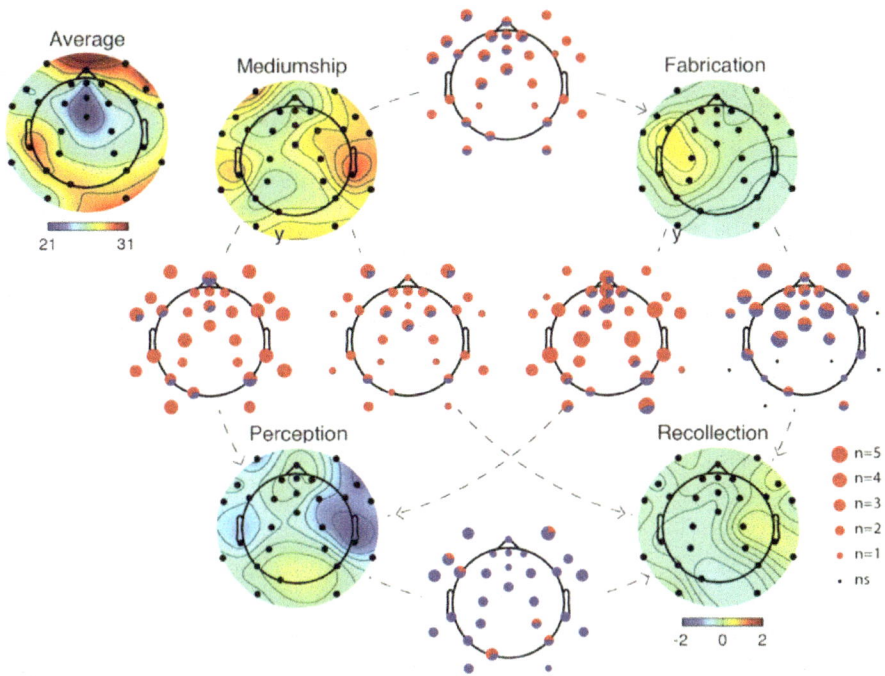

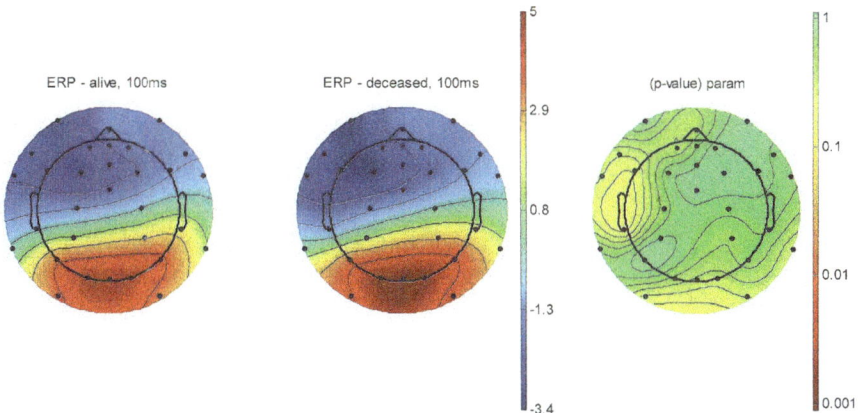

ALIVE   DEAD   ENERGY

Here are the computer diagrams of my own head, generated in November 2014 using Dr. Arnaud Delorme's program for analyzing mediums and meditators. Each circle shows my head. My nose is the triangle at the top, my ears are shown on each side, and the back of my head is on the bottom. At the far left is a diagram when I am observing a photo of a person who is still living. In the middle is a diagram of my head when I am observing a photo of a person who is deceased.

Notice, at the back of my head (bottom of each image), which is where the visual cortex is located, how the diagram on the left is dark red, but the diagram in the middle is a darker red. These are the differences in cognition between "dead" and "alive" images.

The green diagram shows where energy is concentrated in my brain. Notice yellow at my left ear, and yellow at the back of my head. This indicates that my visual cortex is registering the image shown on the computer screen in front of me and the "answer," whether the person in the photo is alive or dead, is being received in my left ear. Of course, that is where I already know Helen hangs out.

But looking back at my own p-value images, my mind appeared to me to be drifting somewhere between FABRICATION, RECOLLECTION, and PERCEPTION, until the scientists pointed out there is no "turquoise" in my p-values, which appears in those other three, and said that this similar appearance was merely coincidental as I transitioned from PERCEPTION to MEDIUMSHIP.

ISOLATION CHAMBER OR CABINET?

Dr. Dean Radin, Ph.D., chief scientist at IONS, took this photo of me just after they had wetted the electrodes on my head, for the thirty-two-point EEG. The towel is to catch the drips. This is just before placing me in the electromagnetic isolation chamber (Faraday cage). I also put my glasses back on for the experiment in the chamber.

Now, back in the 1890s, the mediums would sit in "cabinets" in order to concentrate their energy. Today, at Arthur Findlay College, we mediums still sit in cabinets when we do trance mediumship. But there were so many fraudulent mediums in the 1890s and early twentieth century that this practice was discouraged.

Today, this isolation chamber at IONS certainly looked to me like a cabinet. After I got inside they closed the door so I was supposedly isolated… But my chief Spirit guide, Helen, whispered in my left ear, *"Alan, what are*

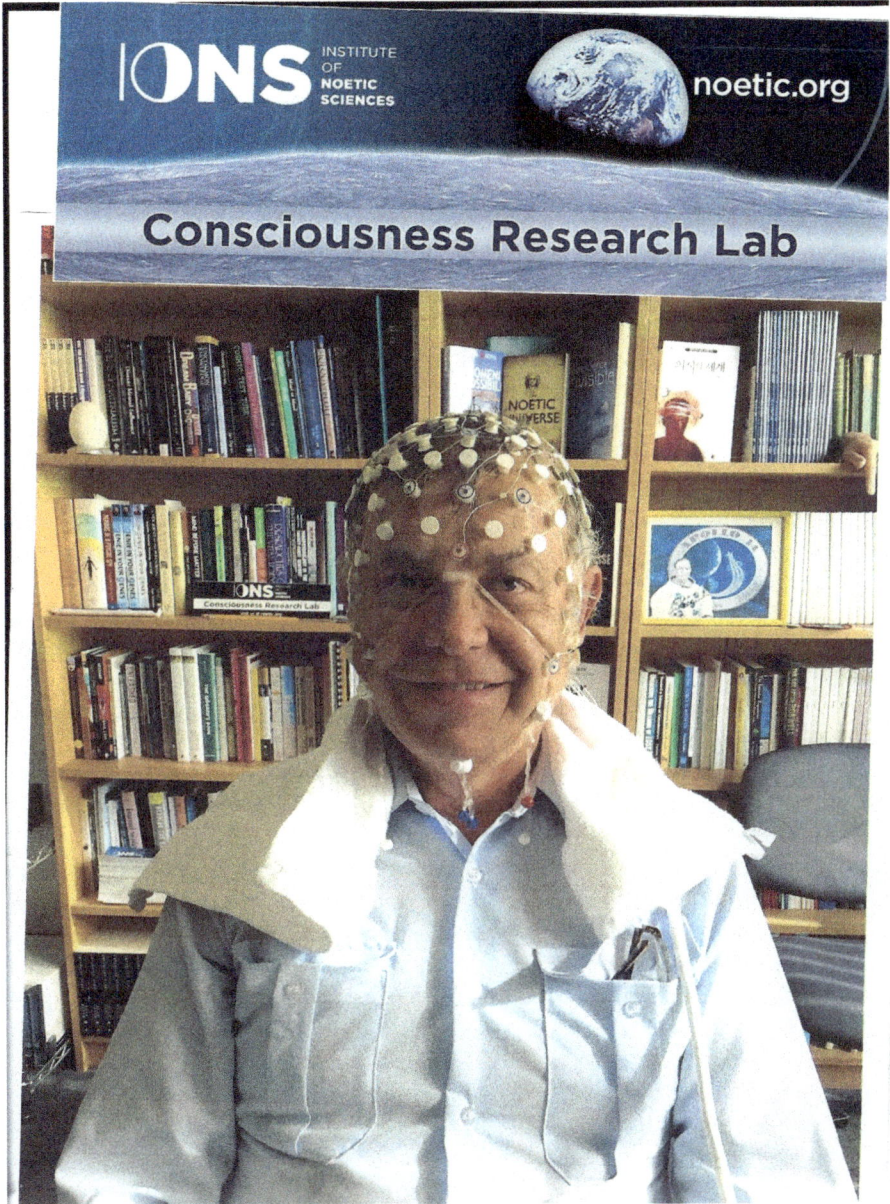

they thinking? I'm right here in the cabinet with you... This is not an isolation chamber...it is a cabinet."

2015 TO PRESENT—WORKING WITH LACH AT UNIVERSITY OF ARIZONA: In the fall of 2015, I began working as a research medium with Dr. Gary Schwartz, Ph.D., director of the Laboratory for Advances in Consciousness and Health (LACH) at the University of Arizona (Lach.web.arizona.edu). While I had been reading Dr. Schwartz's work for over ten years, we had never had occasion to meet formally until June 2015. We were both speakers at a convention of the Academy of Spiritual and Consciousness Studies, and I met both him and his wife, Rhonda, who also spoke at the conference. Subsequently, we have since begun to do significant research together, with me traveling to Tucson and Dr. Schwartz traveling to San Francisco.

Previously, Dr. Schwartz had been studying mediumship for over thirty years. And as a part of an extended research project looking at the controls and accuracy of mediumship experiments, including double and triple blinding, and published in several papers in 2002, 2005, and 2011, Dr. Schwartz found that:

> *"The totality of the experiments effectively rules out potential conventional psychological explanations of (1) fraud, (2) 'cold reading' techniques used by fake mediums (psychic entertainers) to coax information from sitters, (3) visual, auditory, and olfactory cues, (4) sitter rater bias, (5) vague, general information, (6) statistical guessing, and (7) experimenter effects.*

> *"The totality of the experiments also essentially rules out one potential anomalous (i.e., paranormal) explanation: the possibility of telepathy (or mind reading) by the medium of the sitter's mind. For example, in numerous experiments research mediums obtained information that the sitter did not know, which was subsequently confirmed by relatives or friends living hundreds or thousands of miles from the sitter and the medium. And in triple-blinded experiments, the experimenter (the "proxy sitter") was blind to the information about the sitter; hence the medium could not have been reading the mind of the proxy sitter (experimenter) to obtain the accurate information they received about the sitter's deceased loved ones."*

One of Dr. Schwartz's studies, which he undertook with Dr. Julie Beischel in 2007, published in *EXPLORE: The Journal of Science and Healing* [3(1), 23-27] in 2007, concludes:

> *"The results suggest that certain mediums can anomalously receive accurate information about deceased individuals. The study design effectively eliminates conventional mechanisms as well as telepathy as explanations for the information reception, but the results cannot distinguish among alternative paranormal hypotheses, such as survival of consciousness (the*

continued existence, separate from the body, of an individual's conscious-
ness or personality after physical death) and super-psi (or super-ESP;
retrieval of information via a psychic channel or quantum field)."

That paper also suggested the best triple-blinding methods for mediumship studies. These were the control conditions that Dr. Delorme and I had adopted for my research work at the Consciousness Research Lab for the Institute of Noetic Sciences in February of 2015, where I was a research medium working on their current study of evidential mediums' accuracy.

For this program I did readings for thirty-two individuals in two days, with spectacular accuracy, which replicated Dr. Schwartz' prior findings, although I can never find out the precise results due to the triple blinding. But more than 10 percent of those thirty-two sitters (none of whom I knew before the readings) have bumped into me at other IONS events later in the year to say, *"OMG, it was a great reading,"* but without telling me the details and so not un-blinding the experiment.

These readings were probably so good because under these triple-blind conditions, my analytical mind could not possibly get involved in any of the methods controlled against. And so with no analysis or guessing possible, this leaves only one place to find the information—from intuition (Spirit).

In January of 2016, I also became a research medium with the Psychical Research Foundation, working with Bryan Williams, director of research (psychicalresearchfoundation.com). This work is still in progress, so I won't discuss it here.

7. *CONCLUSION:*
EVIDENTIAL MEDIUMSHIP TAKES DEDICATION: My personal experience of mediumship is that anyone can learn to channel spirits from the other side, **if they will only practice long enough.** I did not start working on the platform until my eighth year of studying mediumship. But like becoming a virtuoso concert pianist, it requires patience and dedication. Excellent evidential mediumship, which consistently brings through names and other WOW evidence, is a lifetime work, but anyone with dedication and persistence (and even nerd scientists like me) can learn to do it well.

Accurate evidential mediumship *requires humility, gratefulness, suppression of the ego, and dedication to continual study.* It is not for the faint-hearted, and once you realize its truth, it becomes compelling. When someone says, "Why are you a medium?" my answer is

"I can't help myself. This work is why I live."

FINALLY, mediumship proves that consciousness survival is a scientific fact.
1. We have a logical possibility.
2. We have a great deal of evidence for it.
3. There is no evidence against it!

Therefore, by our scientific requirement for **proof beyond a reasonable doubt**, it is indeed an established scientific fact that consciousness survives and communicates with us.

Three videos that explain what we are doing at the Consciousness Research Lab (CRL) and at the Laboratory for Advances in Consciousness & Health (LACH) are listed below:

- **Dr. Dean Radin discusses the work of the Consciousness Research Lab** (at noetic.org). VIDEO CLIP (2 minutes) at http://vimeo.com/113981492
- *Dr. Arnaud Delorme discusses the MEDIUMSHIP SCIENCE we have been working on together at Noetic.org. VIDEO CLIP (2 minutes) http://vimeo.com/113345358*
- **Dr. Gary Schwartz discusses the Mystery of Consciousness:** Published on Mar 8, 2015. Dr. Gary Schwartz, professor of psychology, medicine, neurology, psychiatry, and surgery at the University of Arizona and director of its Laboratory for Advances in Consciousness and Health. We asked Dr. Gary Schwartz for his thoughts on the mystery of consciousness. https://youtu.be/b1OeVvZCKVQ

8. HOW IS SCIENCE FUNDED, and WHY IS NO FEDERAL MONEY REACHING PARAPSYCHOLOGICAL RESEARCH?

Most scientific research is funded by the National Institute of Health (NIH) or the National Science Foundation (NSF) through federal grants, but none of that federal money is put into research of the paranormal.

This is true even though the most pressing questions, like *WHY ARE WE HERE?*, need answers, these are considered to be "religious" questions. Also, the National Research Council's (NRC) report (published in 1988) states unequivocally that that there was *"NO SCIENTIFIC JUSTIFICATION for any PARAPSYCHOLOGICAL PHENOMENA."*

So, even though the Defense Intelligence Agency for twenty-four years funded the Stargate project (see Chapter 10) with excellent results, and the DIA has an official *1986 Army Remote Viewing Manual* based on that research, the official (disinformation) policy seems to be that there is no scientific reason to look into the paranormal.

> (PDF of the Army manual is available at:
> www.firedocs.com/remoteviewing/.../crvmanual/CRVManual_Fire docsR...and there is an online indexed version posted at:
> www.rviewer.com/crvmanual/)

The Army uses the manual to train intelligence agents in how to do remote viewing (clairvoyance) and become psychic spies, and which was declassified in the late 1990s. Yet the official position of the government is still that there is nothing to it.

Consequently, due to the NRC's official position, all funding for parapsychological research comes from private donations. So please help... Your contribution, even of only $25 to $100, can make a real difference.

THE INTELLIGENT SOLUTION:

TAX-DEDUCTIBLE, SUPERVISED RESEARCH FOUNDATION AT IANDS - San Francisco:

If you wish to support all of these ongoing research efforts I have been discussing, while making one deductible contribution, and also have the assurance that your funds will be carefully supervised (not given to quack research efforts), you can achieve this by contributing through the Research Foundation of the International Association for Near-Death Studies—San Francisco (at IANDS-SF.org), an affiliate non-profit association of the larger international corporation (IANDS.org), founded in 1978. Both organizations are federally approved educational organizations under IRS Code section 501(3)C, and all donations to IANDS-SF are tax deductible. Further, as director of development for both the National Board of IANDS, and as director of the Research Foundation at IANDS-SF, I supervise all these funds, 100 percent of which are distributed to credentialed parapsychological scientists working on consciousness research.

A SIMPLE SOLUTION: What IANDS-SF accomplishes is to allow individuals to make tax deductible contributions to IANDS-SF as a 501(3)C non-profit, which in turn hands out those funds as research grants to consciousness researchers. The simple fact is that most research scientists are not set up as non-profit organizations. Many are post-doc fellows at universities (starving grad students). But the IANDS-SF Research Foundation Committee has the experience and expertise to accomplish things that most donors can't, for example:

- Due diligence required on the grant requests to determine that the proposed experiments will actually be scientific;
- Project surveillance, checking on the operation of the funded experiments, insuring that adequate controls are maintained under clinical conditions (i.e., using Gantz field research methods and double and triple blinding);
- Reviewing progress reports before disbursing further funding; and
- Analyzing the completed study's reports and publications.

DO IT NOW:

As IANDS director of development, I personally want to ask you to join me with a contribution to support this great work. What can you do?

- Can you give $250, $500, a thousand?
- Can you commit to a continuing donation every year?
- Can you donate a set amount each month by credit card?

Go to the website at IANDS-SF.org and simply click the donation button, or email me at Alan@iands.org and I'll telephone you to make other arrangements to receive your contribution.

Alternatively, donations can also be made directly to the following organizations:

International Association for Near-Death Studies (IANDS.org)
2741 Campus Walk Avenue, Building 500 (Rhine Center)
Durham, NC 27705

Laboratory for Advances in Consciousness and Health (LACH)
(Lach.web.arizona.edu)
Department of Psychology, The University of Arizona
1601 N. Tucson Blvd., Medical Square Suite 17
Tucson, AZ 85716

Academy of Spiritual and Consciousness Studies (ASCSi.org)
P.O. Box 84, Loxahatchee, FL 33470

Institute of Noetic Sciences (IONS at noetic.org)
101 San Antonio Road, Petaluma, CA 94592

Note: Helen (who works with me on the other side) while still living in the physical had joined IONS back in the early 1980s. I personally joined IONS in the late 1990s. Today we both work with all these groups.

CHAPTER 9

SCIENCE OF PAST-LIFE REGRESSION THERAPY (PLR) or HYPNOTHERAPY

Another field of research into past lives, which is separate from the work on *Children Who Remember Past Lives* at University of Virginia, is the field of psychology known as *Past-Life Regression Therapy*. Honestly, this is still a fringe area of psychology, and most of my mainstream psychiatrist friends schooled in the Freudian tradition are wary of anyone practicing in this field. My Jungian psychologist friends, while a little more welcoming, are also wary.

There are many professionals now working in this field, and one that I like particularly is Dr. Michael Newton, a psychologist who has done significant work mapping the area between lives as outlined in his 2014 book *Destiny of Souls*. Another is a European psychiatrist, Dr. Marcel Westlund (Swedish, but now living in the Bahamas), who has also done significant work documenting those past lives and has written a 2015 book, *AFTERLIFE: A Psychiatrist Exploring Life Beyond Death.*

THE WORK OF DR. BRIAN WEISS:

One prominent M.D. psychiatrist who has all the American Medical Association Freudian credentials that American Materialist doctors can't argue against, and who is also a nationally recognized expert in parapsychology, but has delved deeply into this controversial area of *past-life regression therapy* is Dr. Brian Weiss, M.D., a practicing psychiatrist and formerly chief of psychiatry at the hospital affiliated with the University of Miami. Dr. Weiss is convinced that reincarnation is real. He describes the problem this way:

> *"I have encountered some extremely talented people—psychics, mediums, healers and others…and I have encountered even more who have limited talent or skill and are mostly opportunists… But I have also been careful not to throw out the baby with the bathwater… One person or one experience might be disappointing, but the next might be truly extraordinary and should not be discounted because of previous events."*

OPPORTUNISTS: Truly, the opportunists abound in the paranormal world. There are fake Tarot readers and fake mediums, and especially of late the fake past-life readers ("Did you know that in a past life you were Napoleon?"). These frauds should be obvious to any careful customer, but they persist. People are just plain gullible and as P.T. Barnum said, *"There is a fool born every two minutes."* And unfortunately, people who don't believe in consciousness survival or karma see no reason not to take advantage of this for personal gain.

But these shyster opportunists are mostly materialists who believe that *since the paranormal is impossible, anyone who works in this field is by definition a charlatan.* They feel that all mediums are liars and fakes and that it won't hurt if they also share in the "stolen goods."

But such opportunists have little idea what damage they have done to the truth, and that this karmic debt will remain in their karma. They also don't realize that true evidential mediums can see right through them to their rotten core. Their *"I am a complete fake"* auras shine out from them like a beacon.

Yet Dr. Weiss is correct, we all must be careful not to throw out the baby with the bathwater. As you will see with Dr. Weiss, the key is to always look at the evidence. Spirit always provides wonderful and undeniable evidence. When you have no verifiable evidence, then you are dealing with a fake. Sometimes, unfortunately, even good-hearted people who work as mediums may not even know they are making it all up. So, the final test of any real medium or past-life regression therapist will be that they have undeniable evidence. One gold standard that I strive mightily to achieve as a medium is to give the name of the unknown loved one. I want that initial evidence to let me know that the message is coming from Spirit and I am not making it up.

DR. WEISS' PROFESSIONAL BACKGROUND: Brian Weiss, M.D. was a disciplined, conservative scientist and physician who distrusted anything that could not be proven by the traditional scientific method. Graduating Phi Beta Kappa, magna cum laude, at Columbia University in 1966, he received his M.D. at Yale University in 1970. Following internship at New York University-Bellevue Medical Center, he completed residency in psychiatry at Yale. Next, he taught at the University of Pittsburgh and the University of Miami, where he led the pharmaceutical division. He became associate professor of psychiatry at the U of M medical school, and chief of psychiatry at the university-affiliated hospital, by which time he had already published thirty-seven scientific papers and book chapters in his specialty field. Weiss is definitely not the type of recognized professional who would lightly espouse fringe ideas.

DR. WEISS' STORY: But in 1980, he met a new patient named Catherine. For eighteen months, Dr. Weiss used conventional methods of therapy, attempting to overcome Catherine's symptoms, but when nothing else worked, he tried using hypnosis. It was during a series of trance states that Catherine began to continually recall her "past-life" memories. It became apparent that she could not remember these past lives at all in her waking state.

In hypnotic trance, Catherine also spoke of the time between lives. It was during one of these sessions while Catherine was in a trance state that **Dr. Weiss was contacted by beings other than Catherine, who called themselves "Masters."** These were separate beings on the other side who were in charge of Catherine's "curriculum." Also, to show Dr. Weiss that they were separate from Catherine, these Masters told Dr. Weiss private things about himself that Catherine could not have known. One of these Masters told Dr. Weiss that he (the Master) had been incarnate in the flesh eighty-six different times. Catherine, while in a trance state, could recall some of what had taken place in her past lives, but *even in trance she could not recall any of the things the Masters had said* to Dr. Weiss through her.

After considerable contemplation, Dr. Weiss decided that regardless of any consequence he might face as a professional doctor, such consequences would not prove to be as devastating as the personal consequences of not sharing the knowledge he had gained about immortality and the meaning of life. So, **in spite of any damage to his own standing in the medical community, he decided to go public with what he had learned.**

Frankly, if a few more scientists were willing to face up to the pseudoscience of the *Materialist* BS, then the shallow-thinking *Materialists* will lose their death grip (excuse the pun) on culture and society.

In his numerous books, among which are *Many Lives, Many Masters, Messages from the Masters,* and *Same Soul Many Bodies,* Dr. Weiss tells his story in great detail and I recommend reading these books.

Basically, Weiss' books tell the same story as Dr. Ian Stevenson's books on reincarnation. Using regressive hypnotic therapy merely bypasses our forgetfulness and allows those of us who can't consciously remember our prior lives to access the information that is dormant in our subconscious. After reading Dr. Weiss' books, it becomes easy to see that reincarnation and having past lives is quite *normal.*

Here are some of the specific things about reincarnation that the Masters told Dr. Weiss (*italics are my inclusions*):

1. **We choose when we will come into our physical state.** (*This agrees with the near-death experience of choosing to come back.*)
2. **We choose when we will leave, because we know when we have accomplished what we came here to do.** (*This agrees with the near-death experience, with the clarification that in the NDE, such decisions are made on the other side of the veil.*)
3. **Everyone's path-goal is the same. We must all learn charity, hope, and faith, and learn them well.** The religious orders (nuns, monks, and friars like the Franciscan and Benedictine orders) of all faiths have come closer to this than the others, because they have given up so much without being asked. The rest of us are always looking for rewards and justifications for our good behavior, but the reward is in the doing and not the getting. (*To understand a Christian mystic's view of this fact, see the prayer of Saint Francis, or read Pierre Teilhard de Chardin.*)
4. **Patience is a virtue; everything comes with timing. Your life cannot be rushed; we must accept what comes to us at any given time, and not ask for more.** We were never really born; we just pass through different phases and there is no end. There are many dimensions, but time is not as we see it. (Consciousness is non-local and non-temporal.) Progress is measured in lessons that are learned.
5. **To be in the physical state is actually an *abnormal* condition.** When you are in the spiritual state, that is *normal* to you. The spiritual state (on the other side) is a state of renewal (R&R); it is a dimension like other dimensions. *This agrees with the near-death experience: home is on the other side, with the Light, while being here in our physical reality is the less desirable state.*
6. **You can waste much energy in fear.** You must stop wasting energy in fear and, instead, release all fear and move forward in love.
7. **If people knew that life is endless, that we will never die, and that we were never really born, then all fears would end.**
8. **Apparently, we cannot learn as spirits because all learning requires feelings, and feelings are inherent in the flesh,** whereas spirits do not "feel." *This agrees with the NDE.*
9. **When we arrive on the other side, we are usually burned out.** We must go through a period of renewal, a period of contemplation, and a period of decision whether to return to the flesh again. Our physical bodies are just vehicles for us while we are in the flesh.
10. **There are seven planes of existence, each one consisting of many levels, and we must pass through all seven before we return to the flesh.** *This closely agrees with the ancient Egyptian religion of Thoth-Hermes, and parallels what Spiritualists have ascertained through after-death communications.*
11. **Karma: We have debts that must be paid.** If we have not paid out these debts, then we must take them into another life in order that

they can be worked through. You progress by paying your debts ("forgive us our debts as we forgive our debtors"). *Plato, Buddhists, Origen, Jerome, and St. Gregory agree with the concept of karma.*

12. **When we are in the flesh, we each have a dominant trait**—lust, greed, etc.—and we each must learn to overcome that dominant trait. If we do not, when we return again, that trait will accompany us into the next lifetime, and the next lifetime will be harder and more complicated. Yet if you overcome your dominant trait, you will find your next life to be easier. You choose it; you are responsible for the life you have. It is *your* karma.

13. **Wisdom is achieved very slowly.** This is because intellectual knowledge is so easily acquired, but is not enough in itself. It must be transformed into emotional or subconscious knowledge. Once it is so transformed, then the imprint is permanent. Theoretical knowledge without practical application is simply not enough for your spirit's evolution.

14. **Understand that no one is greater than any other.**

A NOTE OF CAUTION: It is important to remember that, although this excellent philosophy agrees loosely with the NDE and the religion of the ancients, the jury is still out regarding the above stated ontology of the Masters until additional replication in double-blind and triple-blind conditions can be acquired.

To date, although several additional psychiatrists have also published similar findings of past-life memories in patients through hypno-regression, none have yet reported similar discussions with these Masters. On the other hand, the above stated worldview does coincide well with the findings of *near-death experiences, after-death communications,* and the *children who remember past lives.* While final scientific replication must perhaps wait, my heart intuitively agrees with Dr. Weiss.

In any case, Dr. Weiss' experiences are great reading; he is eloquent, compelling, and an honest guy who clearly says, **"The afterlife is very real."** As quoted below from his excellent book *Messages from the Masters*:

"I believe we do reincarnate, until we learn our lessons and graduate, and I have repeatedly pointed out that there is much historical and clinical evidence that reincarnation is a reality.

"I have become aware that an entire spiritual philosophy has been gently unfolded and handed to me... Our consciousness has finally evolved into accepting this filtered wisdom of the ages... We are swimming in a sea of New Age, holistic, and spiritual awareness that seems to have flooded over the dams of old beliefs and constricted consciousness...the evidence is everywhere that New Thought is becoming mainstream."

I was first personally privileged to hear Dr. Brian Weiss speak in San Francisco in February 2002, and I spoke with him afterward. During his presentation he mentioned several additional cases that had recently come to his attention, which I later researched and I believe you will find interesting in the story below. I heard Dr. Brian Weiss speak again in October 2015.

JENNY COCKELL / MARY SUTTON:[64] This is a great read, and Jenny's book is available from Amazon.com. Jenny is an English housewife with two children. Like Dr. Ian Stevenson's subjects, she spontaneously remembered, beginning in early childhood, that she had lived a prior life in Ireland, where she had eight children, but that she had died before those children were fully grown.

From childhood on, Jenny drew pictures of her house and a map of the shoreline and her church, which had an unusual façade. She researched Irish maps, looking for the town of Malaheit. After she found it, she traveled there and felt that it was the place where she had lived before. Jenny found both the church and her former home. The home had become a ruin and had been empty since the 1950s. But she believed she had lived there in the 1920s or '30s, and that she had been named Mary Sutton. She also believed that she had died from the complications of childbirth, leaving behind her children. So, she kept inquiring.

Today, Jenny has managed to reunite with five of Mary Sutton's eight children who are still living, and she has also found the room in a Dublin hospital where Mary (her former self) died. Four of the children (all older than Jenny is today) now believe that Jenny is their deceased mother, Mary, reincarnated because of all the things she can remember about their lives together. The fifth child prefers to believe that his deceased mother, Mary, speaks to Jenny from another dimension, telling details to Jenny. But Jenny emphatically tells that child of Mary's, *"No, I am actually Mary."*

OTHER LIVES, OTHER SELVES: Another educated scientist who is speaking out is an Oxford-trained Jungian psychologist, Roger J. Woolger, Ph.D., currently living in Burlington, Vermont, who believes fully in reincarnation through his own experiences with past-life regression therapy. His seminal book, *Other Lives, Other Selves*[65], published in 1988, has a marvelous first chapter chronicling his own conversion from skeptic to believer. The book goes on to differentiate between individual past lives and a more Jungian view of karmic collective consciousness or **"shared" past lives**, which casts past-life memories in a very different light, and could explain how several people alive today could remember the same past life. Maybe, all seven people who remember being Napoleon actually were previously Napoleon, or maybe Napoleon had a split personality and each of them is just one of those. The endless possibilities boggle the mind; yet, it is obvious that we know *something* about reincarnation, but not all.

CONCLUSION:

You can look more deeply into the referenced literature in the endnotes and find additional evidence for reincarnation. On the other hand, to date I have personally found no evidence at all for the opposing view: that reincarnation is not valid.

There have been no near-death survivors saying, "There is nothing over there," and no one to date has reported remembering a prior nonexistence. So, not only is there a preponderance of evidence, but all the evidence is entirely on the side of reincarnation. The fact is there are actually hundreds of rigorously documented white ravens.

The near-death experience itself is a form of reincarnation. The consciousness simply left the body and then reincarnated in the same body. This is not very different from reincarnating in a new body.

Again all of this supports biocentrism.

So, then is reincarnation a scientific fact?
1. We have a logical possibility…
2. We have a great deal of evidence for it:
   A. From the children who remember past lives, and
   B. Also from adults who remember past lives through hypnotherapy.
3. There is no evidence against it.

Although there may be a lot of fraud and charlatanism in the genre causing a good deal of doubt, still by our scientific requirement for **proof beyond a reasonable doubt,** it is now a twice-established scientific fact that past lives are real and that apparently we do reincarnate.

Yet in the end, science has not given us any answers as to how this happens. Just how do the bits of information that were "us" actually reincarnate? Do we clone a new "self" or does the same "self" move on? Or do only small aspects (bits) of us move on?

The best answer I have found comes from Suzanne Giesemann (SuzanneGiesemann.com) in a posting of what she received from her Spirit guides (Sanaya) on the morning of May 24, 2015:

> **WHISPY:** The question is so often asked, "Do we reincarnate?" And the answer is a qualified yes. Aspects of our personality do appear to reincarnate, yes. The real you has no form. The human thinks in terms of identifiable, separate forms. That form-like aspect of you does not come back again so that you would recognize it walking

down the street. Yes, we see you smiling at this silly concept. And so we smile as well when you worry that one you love who has passed will reincarnate before you get to join them and so will be lost to you. You need not worry (about anything, ever). Consciousness is an energy form, and it is true what you are taught: Energy cannot be created or destroyed, merely transformed.

You and the one who has died are Aspects of Consciousness ... Aspects of Greater Consciousness. When in human form, the Aspect of Consciousness takes on certain characteristics which result in the soul's growth. This is the point of human existence: adding to the Whole. After you have added what we hope is more Love to the Whole, a bit of the Whole of You may decide to go back and take on form yet again for yet more growth. Those aspects that were you do not completely gel into one form and become lost to those who once loved you. No, the Aspect of Consciousness that was you will always, yes, always remain recognizable as the Aspect of Consciousness known as you, even when bits (if formless energy can be referred to as bits) of the Greater You are now off having another experience as a human for the growth of it.

Do you see? Fear not. There is only Love, only Being, only One Mind, and billions of aspects of That. Love never dies. It merely goes round and round, growing ever brighter, ever stronger as you, and you, and you, and you.

- **Dr. Brian Weiss on Connecting with Your Everyday Angels |** Oprah Winfrey Published on Jun 2, 2013. Past-life regression expert Dr. Brian Weiss says the people who have loved you on this earth are called master teachers—and can be viewed as angels after they pass. Watch as he and Oprah discuss the many forms of angels and why Oprah says she feels more connected to her late dog, Sophie, now than she ever did before. https://youtu.be/GUfjZzm2FFE

- **Brian Weiss: Past-Life Regression Session** Published on Nov 19, 2012
  Author of the best-selling *Many Lives, Many Masters*, Brian Weiss, M.D. has a lifetime of work that reveals the very real physical, emotional, and spiritual transformation that is possible when we embrace reincarnation. In this thirty-minute video, he guides you through a past-life regression experience from the comforts of your home. https://youtu.be/xTnAqDPBsoY
  Explore more from Brian Weiss: http://eomega.org/workshops/teachers/...

CHAPTER 10

SCIENCE OF REMOTE VIEWING

REMOTE VIEWING: Although normally all funding for paranormal research comes from private donations, for twenty-four years the federal government fully funded research into remote viewing. This began with NASA and continued with secret black box funding from various defense department intelligence agencies. Dr. Harold (Hal) Puthoff, Ph.D. and Russell Targ (whom I know through the I-ASC.org) spent many years at Stanford Research Institute (SRI).[66] Also, my colleague and friend, Dr. Dean Radin, Ph.D., chief scientist at the Institute for Noetic Sciences (IONS) with whom I now study mediumship scientifically at IONS, worked on these "Stargate" projects with Puthoff and Targ for several separate agencies of the Department of Defense. Some of the secret records were partly declassified in the late 1990s.

First, these scientists measured and documented remote viewing using double-blind studies to determine whether this phenomenon actually existed. Russell Targ, who now teaches remote viewing methods, calls this quarter century of work the "Remote Viewing Projects" and has written several books on this demonstrating their statistical success which proves that clairvoyance (remote viewing) as a normal human power is valid.

Elizabeth Mayer[67] also described this remote viewing work in great detail, including interviews with Dr. Harold Puthoff, Ph.D., who originally supervised that remote viewing project at SRI, which was funded by the CIA and DIA. Much of their research is still classified. However, the following cases are a couple of examples of their findings that have been declassified. Basically, their research verified that the ESP of remote viewing, or what mediums refer to as clairvoyance, is indeed a demonstrable reality. In other words, mediumship (clairvoyance) is real and can be learned by anyone wishing to follow the protocols developed for the army by Stanford Research Institute.

THE PAT PRICE CASE (July 1974): The CIA gave Hal Puthoff geographic coordinates of a location in West Virginia and asked if Stanford Research

143

Institute (SRI) out in California (3,000 miles away) could tell them what was located there. Puthoff assigned the project to Pat Price. Price was a retired police detective who had used psychic powers to solve cases for years, and had responded to SRI's advertisements looking for "sensitive" people capable of remote viewing. Pat accepted the assignment and immediately sent in a five-page report describing a few log cabins and a couple of roads located at the specified geographic site. That was all he put in the report regarding the geographical coordinates given in the assignment.

But Pat also added, "Oh, by the way, over the ridge there is this really interesting place. That must be the place you are actually interested in." He then went on to describe a secret military site he considered as being highly sensitive with the heaviest security. He provided code names related to the game of pool, along with other information about what was going on there and the personnel involved.

When the CIA got the report, their primary reaction was that it was way off and that the log cabins were merely the location of an Office of Scientific Intelligence (OSI) officer's vacation cabin. But then a few days later, when the OSI officer went out to the site, he discovered that just over the ridge there *was* a highly sensitive underground government installation. The site was top secret and its existence totally a surprise to the CIA. Some of the details in Price's account of the site were wrong, but a stunning number were right, including the code names relating to the game of pool that were used to identify files in a locked cabinet within the facility.

These kinds of spectacular results cause a lot of interest in the military intelligence field, and Lt. Fredrick Atwater of the 902nd Military Intelligence Group at Ft. Meade, Maryland, was drawn to these ideas. Lt. Atwater read a report published by William Braud[68], associate professor of psychology at University of Houston, describing "psi conducive states." Atwater made a list of these traits and then combed through the personnel records of intelligence workers, looking for people with these psi traits. He came up with Joe McMoneagle, who was working as a senior projects officer for the U.S. Army Intelligence and Security Command (INSCOM). After several interviews with Hal Puthoff and Russell Targ, McMoneagle was assigned by the Army to the SRI Remote Viewing Project.

THE JOE McMONEAGLE CASE (1979)—TYPHOON CLASS SUBMARINE: McMoneagle joined the SRI Remote Viewing Project in 1978 and stayed for eighteen years. Here is Hal Puthoff's description of McMoneagle's work quoted from Elizabeth Mayer's book[69]:

"He'd produce masses of data that were really hot and totally inexplicable by ordinary means. One example that had particular

impact on me was when Joe identified that the Russians were building a new form of submarine. Not only were the size and design
judged by our military to be completely impossible. Worse was the
fact that Joe said the Russians were building this huge submarine in
the dead of a frozen Russian landscape with no direct access to
water, so there would be no way to launch it. The whole thing
seemed not just unlikely but crazy.

"It was during the fall of 1979 and it was one of the first operational
targets we received once Joe was on board. A high-ranking naval
officer from the Naval Security Council (NSC) brought us a photo
of a massive, industrial-type building, some distance from a large
body of water, located somewhere in Russia. The U.S. Government
did not know what the building was, what it was used for, or what its
strategic importance was. It was unusual for its size and appeared to
house a good deal of activity so they wanted to know more.

"We gave Joe the geographic coordinates, nothing else. His immediate response was that they identified a very cold wasteland with an
extremely large industrial-looking building that had enormous
smokestacks, not far from a sea covered with a thick cap of ice.
Later, we found out the location was Severodvinsk on the White Sea.

"Since the first quick impression corresponded very closely to the
photograph, we showed Joe the picture and asked what might be
going on inside it. Here's his own retrospective account of the viewing:

"'I spent some time relaxing and emptying my mind. Then with my
eyes closed, I imagined myself drifting down into the building, passing downward through its roof. What I found was mind-blowing.
The building was easily the size of two or three huge shopping cen-
ters, all under a single roof...

"'In giant bays between the walls were what looked like cigars of different sizes sitting in gigantic racks... Thick masses of scaffolding
and interlocking steel pipes were everywhere. Within these were
what appeared to be two huge cylinders being welded side-to-side,
and I had an overwhelming sense that this was a submarine, a really
big one, with twin hulls.

"'What I didn't know was that my session was reported back to the
NSC and created some dissension. The almost unanimous belief at
the time, by all the intelligence collection agencies operating
against the building, was that the Soviets were constructing a brand-

new type of assault ship—a troop carrier, and possibly one with heli-copter capability. A submarine was out of the question.

"'On my second visit, I got up very close … Hovering beside it, I guessed it to be about twice the length of an American football field and nearly seventy feet in width, and at least six or seven floors high (if it were sitting next to a standard apartment building). It was clearly constructed of two huge elongated tubes run side-by-side for almost its entire length. (I didn't think this was possible with sub-marines.) I moved up over the deck and was surprised to see that it had canted missile tubes running side-by-side. This was critically important because this indicated that it had the capacity to fire while on the move rather than having to stand still in the water, which made it a very dangerous type of submarine…

"'After this session I did a very detailed drawing of the submarine, adding dimensions, as well as noting the slanted tubes, indicating eighteen to twenty in all. This material, along with the typed tran-script of my session…was forwarded to…the NSC…

"'We soon received a follow on request…to return to the target and try to provide an estimated time of completion…

"'I revisited the site and, based on the speed of construction and the differences in the condition of the submarine from one session to the next, I guessed that it would be ready for launch about four months later—that would be sometime in the month of January (1980). A singularly crazy time of year to launch a submarine from a building not connected to water, near a sea frozen over with ice yards thick. (I reported that very soon a crew of bulldozers and other types of heavy equipment would arrive to cut a channel lead-ing to the sea).'"

Reconnaissance photos taken by US satellite in January 1980 showed a new canal running alongside the building and out to the White Sea. Also in the photos was clearly a two-hulled submarine with twenty missile tubes. This was the first of the Typhoon class of submarine, the largest submarines ever built. And it was exactly what Joe McMoneagle had described. There are many books on the market that describe the SRI remote viewing projects in great detail, especially now that about half of the top secret records have been declassified. It is really a hot topic among people on the cutting edge of scientific exploration. If you want to know more about remote viewing, I do recommend reading Elizabeth Mayer's book *Extraordinary Knowing*, which covers all of this in greater detail. Remote viewing is merely a process of getting the "consciousness channel" tuned with the "will channel" within an individual's mind. What McMoneagle called "shutting down my conscious mind" is this turning down of the noise in Walker's consciousness channel, in order to access through the p-car established by the two channels all the shared fragments of mind experience that have been added to the earth's limitless data storage system (Akashic record).

The military defines remote viewing to be the same as clairvoyance. As a medium, I see very little difference. Any way you want to slice it, ESP is real whether we call it mediumship, clairvoyance, remote viewing, mind reading, or psychic reading. And consciousness operating outside the body is real. Again all of this supports biocentrism.

CONCLUSION: So, then is consciousness outside the body a scientific fact?
1. We have a logical possibility…
2. We have a great deal of evidence for it, and
3. There is no evidence against it.

Therefore, by our scientific requirement for ***proof beyond a reasonable doubt,*** remote viewing (clairvoyance) is already an established/accepted scientific fact. The only thing stopping recognition of ESP as a fact by the scientific establishment is its own fully unrecognized religious superstitions and materialist BS.

But the proof of remote viewing shows that what mediums do is a normal function of consciousness. Unfortunately, the National Research Council's (NRC) report (published in 1988) had no knowledge of all this top secret work at SRI, and so it states unequivocally that there is "NO SCIENTIFIC JUSTIFICATION for any PARAPSYCHOLOGICAL PHENOMENA." The official position of the government is still that there is nothing to it.

- **Remote Viewing with Russell Targ—Posted Jan 2011** What do the healer, the mystic, the psychic, and the spy all have in common? They are all in touch with their non-local mind and our community of spirit. During the 1970s and 1980s, Stanford Research Institute (SRI) carried out investigations of our ability to experience and describe distant events blocked from ordinary perception. This intuitive capacity was named "remote viewing." The research was supported by the CIA and many other government organizations for gathering intelligence about worldwide activities during the Cold War. https://youtu.be/IH5_X cq8EnM

- **Testing Joseph McMoneagle - The Remote Viewer** Joseph McMoneagle (born January 10, 1946 in Miami, Florida) was involved in remote viewing experiments conducted by U.S. Army Intelligence and the Stanford Research Institute. He was one of the original officers recruited for the top-secret program now known as the Stargate Project. Along with Ingo

Swann, McMoneagle is best known for claims surrounding the investigation of remote viewing and the use of paranormal abilities for military intelligence gathering.
https://youtu.be/Nf5BmV3HOFI
- The DIA's 1986 *Remote Viewing Manual,* declassified in the 1990s, is available to the public at:
www.firedocs.com/remoteviewing/.../crvmanual/CRVManual_FiredocsR...
Online there is an indexed version posted at:
www.rviewer.com/crvmanual/

CHAPTER 11

SCIENCE OF QUANTUM ELECTRODYNAMICS

"Anyone who is not shocked by quantum theory has not understood it."
—Neils Bohr (1885–1962)

"In light of these considerations, strong doubts about personality survival, based solely on the belief that postmortem survival is incompatible with the laws of physics, are unfounded.[70] —Henry Stapp

Most of us are content to know that quantum physics is true, but aren't interested in understanding the mathematics behind it all, so we don't study it any further. Physics doctoral students are simply happy to have mastered the complicated mathematics, so they don't study it any further either. Yet what Henry Stapp is saying is that there is no conflict between consciousness survival in an afterlife and the physics of the subatomic world as expressed through quantum electrodynamics (QED). And this message is being ignored by shallow-thinking *Materialists*.

On the other hand, when this message is understood, all the phenomena currently called "paranormal" are actually more "normal" and consistent with QED than are the First Principles of classical Newtonian *Materialism*. At that point the shift in perspective causes *Materialism* to suddenly become abnormal.

This new picture of a conscious universe that is emerging shows an interconnection between all things filled with mind-matter interactions and instant communication across vast distances.

Dr. Erwin Laszlo, in his latest book, *New Concepts of Matter, Life and Mind*, writes that:

"The living world is not the harsh domain of classical Darwinism, where each struggles against all, with every species, every organism and every gene competing for advantage against every other. Organisms are not skin-enclosed selfish entities, and competition is never unfettered. Life evolves, as does the universe itself, in a 'sacred dance' with an underlying field.

> *This makes living beings into elements in a vast network of intimate relations that embraces the entire biosphere itself an interconnected element within the wider connections that reach into the cosmos.* "[71]

To make understanding this as painless as possible, I hope to have reduced this to a handful of postulates so that when you take the time to grasp each postulate individually, you will develop the intuition that allows this to suddenly all make sense.

THE MATERIALIST PARADIGM: First, let us redefine the opposing viewpoint, which is the *Materialist* paradigm, as more fully described earlier in Chapter 4. This is the viewpoint we are all taught in high school science classes, that there is a solid physical reality "out there" made entirely of "unconscious," inert matter. And since the days of Aristotle, philosophers and scientists have been breaking that matter down into smaller and smaller "solid" particles. For 2,300 years, from Aristotle (384–322 BCE) until the twentieth century, science expected that the smallest solid particle would be the atom. It was only in the early twentieth century that we discovered atoms were made of yet smaller particles.

QUARKS, STRING THEORY, SUPER STRING THEORY: The science of QED found that the atom was made up of electrons, protons, and neutrons, but then subatomic particle physics progressed further, finding that these were also not the smallest particles. Instead, the baryons (for example, protons and neutrons) were made of even smaller particles. The quark model was independently proposed by physicists Murray Gell-Mann and George Zweig in **1964**, and soon at the Stanford Linear Accelerator, experiments were showing that quarks had six flavors (up, down, strange, charm, top, and bottom). Finally, I remember celebrating in **1995** with a physicist friend at University of Washington when the top quark was the last to be discovered at Fermilab.

But interestingly, a hundred years before the discovery of the quark, clear back in 1895, a pair of Spiritualist mediums (Charles Ledbetter and Anne Besant) who had been doing "occult chemistry" through clairvoyance (or remote viewing of the interior of atoms) published their findings in the *Theosophist* journal. Not surprisingly, their account of the force binding together the fundamental constituents of matter agreed with our present day string theory model, the super string theory model, and the quark model. These mediums did not run experimental tests; they just meditated, went inside, and looked at the makeup of the atom.[72]

Then, after discovering quarks in the mid-1980s and '90s, quantum particle physicists made the further amazing discovery that these elementary particles were made of ***nothing that was physical.***

The way we explain this "being made of nothing physical" is to say that the smallest particles are like a vibrating string, but without any string being there to vibrate; instead it is just the vibration itself. Radio science has another way to say "just vibration" and that is "frequency." We speak of radio waves having a certain frequency. Using this terminology we could say that the smallest elementary particles are just a frequency, or as Erwin Schrödinger would describe it, "just a wave."

Interestingly, Spiritualists have always spoken of the spirits (surviving consciousnesses) as being at a difference frequency (vibration) than the physical.

Quantum mechanics (later to be redefined as quantum electrodynamics or QED) began to be developed as a theory in 1900 when Max Planck used the word "quanta" to describe the units of force that kept subatomic particles together. QED includes the discoveries of Albert Einstein, <u>Erwin Schrödinger</u>, Werner Heisenberg, Neils Bohr, Wolfgang Pauli, and Paul Dirac, and the theory they advanced in the 1920s and '30s is still being extended today with the Large Hadron Collider at CERN.

MATERIAL REALISM: In the *Materialist* worldview everything is made of matter (substance); thus consciousness and all subjective phenomena related to it, such as conscience, are relegated to being mere epiphenomena (an accidental byproduct) of matter, and therefore are without any causal efficacy (without any real purpose). But that is the same as to say *your consciousness has no purpose.*

When we think of consciousness this way, as merely being an epiphenomenon (byproduct) of the brain, as *Materialists* are forced by their dogma to do, then there is a paradox. Scientists call it the "quantum measurement paradox."

Simply stated, the quantum measurement paradox (or the "measurement problem") is this truth discovered in QED:
Since matter only has potential (in other words exists as a wave) until it is observed by a living consciousness (measured), then matter is forced to remain in the "undefined" quantum state of pure potential until a living consciousness observes it.

Obviously then, if matter is merely a "potential" until a consciousness observes it, then matter is the epiphenomenon (byproduct) of consciousness, and not vice versa. *OMG, who said that? Kill the infidel!* This heresy is simply too much for any *Materialist* believer, so they vehemently attack anyone intelligent and brave enough to say this.

Consequently, since QED shows that the wave can only precipitate into reality as matter after a living consciousness observes it. Then the question comes up, "just what consciousness was doing that observing back when the big bang started the universe? According to QED, there had to be a living consciousness observing in order to precipitate the matter of our universe before it could precipitate as physical reality in the "big bang," and just who, or what, was that living conscious observer?

Oh my God (excuse the pun), but did we just allow Materialist science to discover God as the observer? Now we really need to kill that infidel...let's write out a contract on his life... We can't allow this to go unpunished. This is how small-minded, shallow-thinking, pseudoscientific *Materialists* actually think. They literally believe that I am a fraud, a liar, and a charlatan, that I never went to the other side during my NDE, that I could not possibly talk to dead people. That everything I have experienced is a delusion. They simply "know" without ever looking outside their box at the real data proving the so-called paranormal...that they already comprehend everything. But how stupid, childish, and arrogant that really is, and how immature their intellects seem to be is beyond them.

Obviously, you can see what a big problem this "allowing of consciousness" truly is for an atheist *Materialist*, and why it sticks in their craw. They pride themselves on not having any superstitions like religion. But QED automatically implies a Universal Consciousness (a god), while showing *Materialism* to be a superstitious religion.

Like Yoda in George Lucas' *Star Wars*, I prefer to remain scientific and call this observer The Force, instead of saying "God."

Following are five postulates, each of which is fully compatible with QED, and which when taken together point to the opposite conclusion from what the *Materialists* shallowly hope to believe. Instead of consciousness being a byproduct matter...matter itself is an emergent property (epiphenomenon) of consciousness... Consciousness came first, which turns *Materialism* on its head. The postulates are listed here and then explained later in this same chapter.

FIRST POSTULATE - *OBSERVATION IMPACTS THE SYSTEM*: It is quantum electrodynamics which first illustrated that a system exists in superposition, that is, in all possible states at once, until we observe it to be only in one specific state. This is clearly illustrated in the double slit experiment, first discovered by Thomas Young in 1801. The double slit clearly sets aside *causality*, *determinism*, and the notion that reality is "out there" as it blurs the line between the observer and the system being observed. But this truth strikes at the heart of the materialist view of a "clockwork" universe operating without consciousness.

SECOND POSTULATE - *EVERYTHING IS COMPOSED OF CONSCIOUS-NESS, NOT MATTER:* The atom is 99.999999999 percent empty of matter, and even the "matter" that is there is not made of substance. The protons, neutrons, and electrons are made entirely of quarks and gluons, which in turn are made entirely of vibrations. But there is no "matter" there doing the vibrating. The basic essence of all things is merely a vibration, which itself is just a concept...a thought...mere consciousness.

In other words, what we think of as solid is really an illusion, and the smallest "particle" is not a particle but is only a thought.

Further, it is perfectly possible for other solid worlds, vibrating at different frequencies, to coexist in the same non-local space as ours, in the same way that different radio signals coexist all around you right now. Turn your cell phone ON and the signal is there in your cell phone in your hand. Turn your cell phone OFF and *the signal is still there in your hand* but your cell phone is just not now receiving it. Similarly, like the radio waves, other dimensions of consciousness can exist in the same non-local space with us. This is where the multi-universe theories are allowable, but not necessarily supported by QED.

THIRD POSTULATE - *TIME IS ONLY A PERCEPTIVE ILLUSION:*
As a medium, I am constantly told by beings in the afterlife that time moves very differently for them and that the higher realms are outside of time entirely. Also, I had this same experience of time not being real, or at least not being the same, during my near-death experience. Prior to the QED this idea of time being a perceptive illusion seemed impossible. But QED clearly shows that this concept of time being an illusion is a fact.

FOURTH POSTULATE - *EVERYTHING IS CONNECTED (QUANTUM ENTANGLEMENT):*
At the subatomic level, everything is interconnected and information moves instantly across vast distances; this is called quantum entanglement. This was proven by John Bell's theorem of *non-locality,* which has been replicated in scientific experiments nine times. This explains how psychic phenomena work and how people have experiences of "direct knowing."

Quantum physicist Erwin Schrödinger said,
"Connectivity is not just a property or reality; it is the property or reality."

Dr. Dean Radin Ph.D. said,
"Connectivity among all things is a basic constituent of the fabric of reality."

Dr. Radin further claims that there is experimental evidence of this when random number generators are influenced by intention and even attention. This leads us to yet a fifth postulate:

FIFTH POSTULATE - *THERE IS ONLY CONSCIOUSNESS*:
And this fifth postulate has staggering implications. Quantum electrodynamics (QED) suggests that at the very deepest level of reality there is no matter, but only consciousness.

Now let's examine each postulate in greater detail.

FIRST POSTULATE - OBSERVATION IMPACTS THE SYSTEM:

"We choose to examine a phenomenon which is impossible, absolutely impossible, to explain in any classical way, and which has in it the heart of quantum mechanics. In reality, it contains the only mystery. We cannot make the mystery go away by 'explaining' how it works. We will just tell you how it works. In telling you how it works we will have told you about the basic peculiarities of all quantum mechanics.

"We would like to emphasize a very important difference between classical and quantum mechanics. We have been talking about the probability that an electron will arrive in a given circumstance. We have implied that in our experimental arrangement (or even in the best possible one) it would be impossible to predict exactly what would happen. We can only predict the odds! This would mean, if it were true, that physics has given up on the problem of trying to predict exactly what will happen in a definite circumstance...

"Yes! physics has given up. We do not know how to predict what would happen in a given circumstance, and we believe now that it is impossible—that the only thing that can be predicted is the probability of different events.[73]
—*Richard Feynman, Feynman Lectures on Physics (C)1964 Cal Tech*

THE ONLY MYSTERY IN PHYSICS: As Richard Feynman said, *"Matter, the substance of our reality, only exists as a potential* **until a consciousness observes it.** *Once it is observed it "precipitates" into matter. But it requires a living consciousness to observe it in order for this to happen.*

HOW CONSCIOUSNESS CREATES REALITY: Many people are not aware that QED, our most proven theory of all time, actually supports the existence of consciousness as a building block in physics. The best illustration of this is the double slit experiment as follows.

A BEAM OF LIGHT CHOOSES TO BE EITHER A PARTICLE (PHOTON) OR A WAVE: If a light beam directed at two slits in a screen chooses to behave as a particle (a photon), it will create the pattern shown in the left diagram below. In that case each photon particle passes only through one slit.

On the other hand, if it chooses to behave as a wave, then it will create the interference pattern shown in the right diagram below.

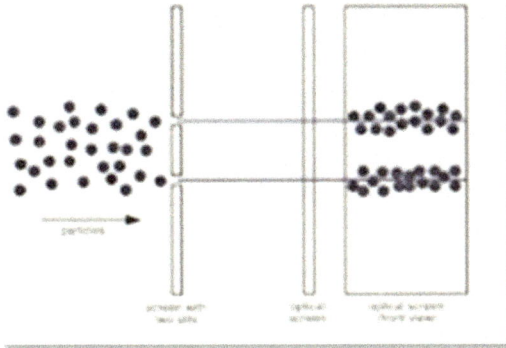

NORMAL PATTERN OF PHOTONS

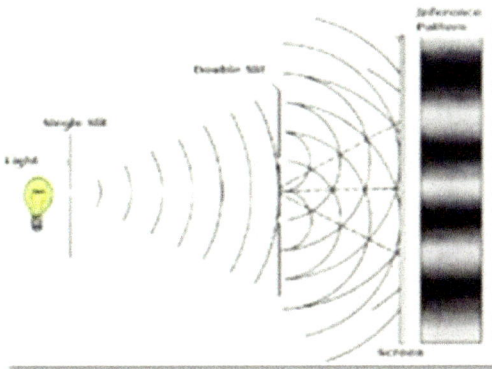

INTERFERENCE PATTERN

Now here is some quantum strangeness: that light beam allows the act of observation to decide which one it will be, either a particle or a wave. The act of consciously observing the event collapsed the wave function, creating the reality of a physical particle. But without that observation it will not become a physical particle. An excellent explanation of this double slit experiment is located online at this URL:

- Dr. Quantum – Double Slit Experiment, Uploaded on Dec 27, 2010. A short lesson in the basics of quantum physics. This BBC doc is perfect: https://www.youtube.com/watch?v=EAv0h or at

- Uploaded Sept 13, 2006: https://www.youtube.com/watch?v=DfPeprQ7oGc

Please take the trouble right now to watch this five-minute video by Dr. Fred Alan Wolf, Ph.D., physicist. It shows the physics behind this phenomenon, which was first discovered by Thomas Young in 1801.

It explains that when we choose to observe the double slit experiment, the light beam decides to act like a particle and only goes through one slit at a time. In other words, THE LIGHT BEAM IS AWARE THAT IT IS BEING OBSERVED and THE OBSERVER COLLAPSES THE WAVE FUNCTION SIMPLY BY OBSERVATION.

If the observation is made by an instrument, not a consciousness, then the light waves remain in a quantum state (wave of potentiality) until the instrument observation is actually witnessed by a consciousness. It takes a living consciousness to make the observation.

Dr. Radin has run specific trials using the double slit apparatus and allowing test subjects (people) to direct their observation to the slits, and then later withdraw their observations of the slits, and the photons or interference patterns match the observation or non-observation. Indeed, observation is what collapses the wave function, with odds against the chance of 184,000 to 1. He published these results in a peer-reviewed journal in 2012. He also found that meditators performed much better with this than non-meditators, at odds of 300,000 to 1, a fact that was predicted by the ancient yogi sage Patanjali more than 2,000 years ago.

OTHER PROOF OF MIND OVER MATTER: Dr. Radin at the Institute of Noetic Sciences (noetic.org) has found that mind controls matter, but Princeton Engineering Anomalies Research (PEAR) has also consistently demonstrated this is true through over fifty million trials during thirty years. PEAR studied the role of consciousness in establishing reality; the data shows machine outputs correlate with the pre-stated operator intentions, and this correlation is statistically replicable—*the machine does what the operator wills it to do.*

After 30 years of continual testing, PEAR has overwhelmingly shown that mind influences matter every time and with odds better than a billion to one.

Mental effort can always control machines. Using volunteers off the street who have no preparation or practice, and asking them to try to influence the outcome of machines, PEAR has shown that these novices tilt the odds in their favor every time!

But this causes a problem because, although consciousness is real, it has never yet been included in the standard model. It is purposefully left out by materialists who, at the same time, are searching for a "Theory of Everything."

WATCH CONSCIOUSNESS AFFECT THE MACHINE:

For Christmas 2012, I got a Psyleron lamp invented by PEAR, which demonstrates mind over matter. Using a random number generator to create multiple colored light, it can be influenced by the observers. You can think it into green, red, blue, yellow, orange, purple, or white. If you watch the video listed on the following URL, you can see my lamp turn light green as the majority of the video crew present just happens to think green all at the same moment, when I say "The lamp will turn green." But then it quickly fades out again. This change to the chosen color accidentally takes place during my September 2013 TV interview with Mel Van Dusen. In the interview the lamp is in the background.

- **"SCIENCE OF THE AFTERLIFE,"** Dr. Alan Hugenot. What it is like when you die? http://www.youtube.com/watch?v=s G8RAVh4VwE

This brings us to the first postulate.

FIRST POSTULATE: OBSERVATION IMPACTS THE SYSTEM

Consciousness is a real force and can affect matter and operates at a distance, OUTSIDE THE BODY, proving that psychokinesis (PK) or "mind over matter" is also real. Consciousness cannot then be a byproduct of the brain, and so may instead pre-exist the evolution of the brain.

SECOND POSTULATE: EVERYTHING IS COMPOSED OF CONSCIOUSNESS, NOT MATTER

"Matter is not made up of matter." —Hans Peter Durr, a longtime coworker with Werner Heisenberg

In the preceding chapters we've covered a lot of facts and we've stated the general theory of a conscious universe. We've also shown that a conscious universe is compatible with QED, where physical materialism is not.

So, it is obviously time to move forward in accepting the scientific data showing that the universe is made of consciousness rather than substance (matter). Here is what physicist Peter Russell Ph.D., said about this in 2001.

> *"Rather than assuming that consciousness somehow arises from the material world, as most scientists do, we need to consider the alternative worldview put forward by many metaphysical and spiritual traditions...in which consciousness is held to be a fundamental component of reality. As fundamental as space, time, and matter, perhaps even more so... When we do, everything changes, and everything remains the same."*

This is true because according to quantum electrodynamics (QED), matter itself is only an illusion fabricated by our collective consciousness and perceived as real and tangible. The celebrated scientist Albert Einstein admitted this when he said,

> *"There is no place in this new kind of physics (quantum mechanics) for the field and (also) matter, for the field is the only reality."*

This brings us to the fact that, if the universe is conscious (made of consciousness) then consciousness survival is the most reasonable explanation for what occurs at dissolution (death of the physical body).

> *"This conclusion that nature is fundamentally mind-like is hardly new. But it arises here not from deep philosophical analysis, or religious insight, but directly from an examination of the casual structure of our basic scientific theory."* —Henry P. Stapp, Ph.D. Lawrence Berkeley Laboratory, University of California, Berkeley, CA

> *"The elaboration of orthodox quantum mechanics (QED) that achieves the most commonsensical solution to the biocentrism problem parallels an elaboration that naturally accommodates personality survival."* —Henry P. Stapp, Ph.D. Lawrence Berkeley Laboratory, University of California, Berkeley, CA

NDES AND CONSCIOUSNESS: Current clinical research is also leading us inexorably to the conclusion that consciousness exists independent of matter, and that every individual's consciousness does, in fact, survive physical death, just as the NDE has been demonstrating for all of recorded history. Recently, Dutch cardiac surgeon Dr. Pim van Lommel, M.D., who has made a study of cardiac arrest NDEs, made the following statement:

> *"Our consciousness is intrinsically connected with non-local space, or the vacuum, (which) is the source of both the physical world and consciousness, while in turn non-local consciousness is the source of both waking consciousness and all other aspects of consciousness... The findings of NDE research suggest that non-local consciousness is present at all times and will therefore last forever."* —Dr. Pim van Lommel, M.D., *Consciousness Beyond Life*, p.307-308, & 317

THE PERCEPTIVE PROBLEM: Classical materialist physics is itself dependent upon the existence of a subjective consciousness, existing entirely outside the perceptive limitations of 3-D plus time, in order to make the empirical observations. But there is nothing in physics, chemistry, biology, or any other classical Newtonian materialist science that can account for an interior worldview (consciousness). The materialist's perceptive problem is that they can't have empirical observations without consciousness, yet they can't explain consciousness.

On the other hand, one of the first objections usually raised against consciousness survival in an afterlife is *"Well, okay, I will admit that there may be a great deal of evidence, but where could the afterlife possibly exist?"*

The short answer is that *the afterlife can easily exist in the dark energy of our conscious universe.*

In **1925,** Erwin Schrödinger derived a linear, partial differential equation that describes how the quantum state of a physical system changes with time. This *Schrödinger wave function* is the most complete description that can be given to any physical system within our universe. Waves don't actually exist as physical reality, but instead exist as potential physical reality, until a living consciousness observes them.

FIRST AXIOM: Matter precipitates from the collapse of the wave function. The collapse of the wave function is a continuous process within quantum electrodynamics, which is instantaneously creating our reality (matter) from our collective perception. Back in **1926,** Max Born (Nobel Prize 1954) showed us that these wave functions defined by Schrödinger were actually waves of probability, not reality.

Building on that, Paul Dirac, in **1928,** fused Schrödinger's equation and Einstein's relativity into a new *Dirac equation* of quantum electrodynamics (QED) (Schrödinger & Dirac shared the 1933 Nobel Prize). Next in **1934,** Werner Heisenberg (1932 Nobel Prize winner) reinterpreted this equation to incorporate Enrico Fermi's theory of beta decay to give us the modern *Dirac field equation,* which expresses a quantum theory of force fields.

This new equation also shows us that particles in the vacuum of non-local space are continually being created and annihilated from the zero-point field of dark energy, every ten septillionths of a second (or every ten yoctoseconds). Without this constant conversion of dark energy into light energy and back again, occurring in just the exactly right quantities, our reality would immediately cease to exist and disappear into a dark hole. And unfortunately, as elegant as the Dirac equation is, it still excludes consideration of the consciousness of the observer in taking an observation, which the double slit experiment clearly proves is taking place.

The **COPENHAGEN INTERPRETATION:** Attempting to define this new paradigm led to the Copenhagen Interpretation, which simply says that "*a particle exists in all states at once until observed.*" This interpretation was created by Niels Bohr, Werner Heisenberg, and Max Born, with strong support from Wolfgang Pauli and John von Neumann.

Yet realizing that the term "observation" as used in the Copenhagen Interpretation is only a euphemism for "consciousness making an observation" places an unanswered question before physicists, and leads inevitably to the incredible conclusion that *mind or consciousness affects matter.*

ORTHODOX COPENHAGEN INTERPRETATION: John von Neumann was able to formalize the mathematics of this Copenhagen Interpretation into a cogent theory. He found that,

> "*If the Dirac field equation describes the true nature of reality, and if Schrödinger's wave function is a complete representation of reality, then the choice made by one's mind in taking the observation has a direct effect on physical reality, regardless of what that consciousness may yet prove to be.*"

Here is how Schrödinger described this new perception:

> "*Inconceivable as it seems to ordinary reason, you, and all other conscious beings, as such, are the all in all. (The soul of the conscious universe.) Hence this life of yours which you are living is not merely a piece of the entire existence, but is in a certain sense the whole... Thus you can throw yourself flat on the ground, stretched out upon Mother Earth, with the certain conviction that you are one with her and she with you. You are as firmly established, as invulnerable as she, indeed a thousand times firmer and more invulnerable. As surely as she will engulf you tomorrow, so surely will she bring you forth anew to new striving and suffering. And not merely 'someday.' Now, today, every day she is bringing you forth, not once but thousands upon thousands of times, just as she engulfs you a thousand times over...*"

So we knew in 1926 that our reality was being continually created by our perception, and that both time and space were merely perceptive illusions, created as a sort of "intellectual dodge," useful in understanding our day-to-day physical existence.

The problem is that most *Materialist* scientists are (as the Gnostics described it 2,000 years ago) *"infatuated with matter,"* so they prefer the nineteenth-century idea of a mechanistic *Materialism* operating an objective reality, which includes real matter. To them the idea that everything is only potential is just too uncomfortable.

CREATING REALITY: The collapse of the Schrödinger wave function, which creates reality, occurs because of an observation. However, because the observation is SUBJECTIVE and not OBJECTIVE, it is not accounted for in any of the equations of quantum mechanics.

In the May 15, 1935 issue of Physical Review, Albert Einstein co-authored a paper with his two postdoctoral research associates at the Institute for Advanced Study, Boris Podolsky and Nathan Rosen. The article was entitled "Can Quantum Mechanical Description of Physical Reality Be Considered Complete?" but it has become known as the EPR corollary. This critique, leveled against the Copenhagen Interpretation, states that **"if interacting systems satisfy separability and locality, then the description of systems provided by state vectors is not complete."**

But seeking a complete description, in 1952 David Bohm inserted this act of SUBJECTIVE OBSERVATION into the Dirac field equation as a factor he called the hidden variables. The EPR correlation had called for this in 1935, and de Broglie had unsuccessfully tried to solve this same paradox with similar hidden variables in 1927. Bohm's insertion of hidden variables remains controversial. But in 2000, Dr. Evan Walker, Ph.D. defined these hidden variables as two separate factors being transmitted by our consciousness and our will, stating how these two hidden variable signals (will channel *and* consciousness channel) *originating from an observer's consciousness disappear as they simultaneously enter the Dirac field equation because the two signals cancel each other out. Yet they also cause the wave function to collapse, thus precipitating the matter of our reality.*

> **"For every action there is an equal and opposite reaction, and the phenomena that are controlled by one's will work exactly the same way. The state that occurs is the state that is willed, and the state that is willed is that state that occurs.**

> *"If the consciousness channel were equal to the will channel, then whatever one wished would also be exactly what happened. Every thought would be a perfect act.*

> *"But the will channel that causes the events that occur is only a small part of the mind. The consciousness channel, which is of quantum mechanical origin just like the will, is much greater in magnitude than the will channel. Our problem with understanding how these phenomena occur arises not with figuring out how the mind can cause a particular thing to occur, but in understanding why only a small part of what we consciously wish for is carried into reality."* —Dr. Evan Harris Walker, Ph.D., Physics Ballistic Research Laboratories of the US Army's Aberdeen Proving Ground, Maryland

According to Dr. Walker, the answer to this question is to be found in the small magnitude of the will channel as compared to the crowded signal in the consciousness channel of the mind. So, in quantum electrodynamics, instead of the fixed state reality materialists hope to find, reality only exists as an infinite number of potentialities (expressed by the Dirac field equation), and remains merely potential until a single potentiality will be precipitated into matter by its being consciously observed (measured). The conscious will of the observer chooses which one of the potentialities will be observed. This choice is known as the state vector collapse, and it is caused by the hidden variables of consciousness and will, inserted as factors into the equation by mental forces of observation by a living consciousness.

Again none of those ways of seeing reality are conclusive ideas from which we cannot deviate, but each helps to provide a context for a new perspective on how it might work.

SECOND AXIOM: *Consciousness provides the hidden variables that precipitate the state vector collapse, forming the matter of our perceived reality. "Mind can control matter."*

> *"I am fully convinced that the soul is indestructible, and that its activity will continue through eternity. It is like the sun, which to our eyes seems to set in the night, but it has in reality only gone to diffuse its light elsewhere."*
> —Wolfgang von Goethe

BUT WHERE DOES CONSCIOUSNESS EXIST? Since consciousness is not energetically discernible within our paradigm of classical Newtonian physics, then where does it actually hang out?

First, to be in full accord with QED and *non-locality,* we need to get rid of any ideas of "where" something exists. There is no space nor dimension and consequently no "where" for it to be located in. *Non-locality* also means "no location." Everything is instead located in potential rather than in a physical space. Everything exists in what we currently call "super-position," which really means "everywhere at once." So, eventually we'll have to replace that word "super-position" with "exists non-locally" or words to that effect.

HOW DOES CONSCIOUSNESS EXIST? It is a better question to ask "HOW" consciousness exists instead of "WHERE." There are two possible ways to "place" the reality of consciousness within our ontological paradigm. One, which has been favored since James Clerk Maxwell first gave us the equations for electromagnetic charge, is out-of-body (or outside of physical matter). This theory postulates that consciousness exists at a higher frequency on the electromagnetic radiation scale.

HIGHER FREQUENCY: Many mediums hold that consciousness survival takes place at higher frequencies on the spectrum of electromagnetic radiation. That matter, at these higher vibrations shown in Figure 12-1 above as

VISIBLE AND INVISIBLE VIBRATIONS

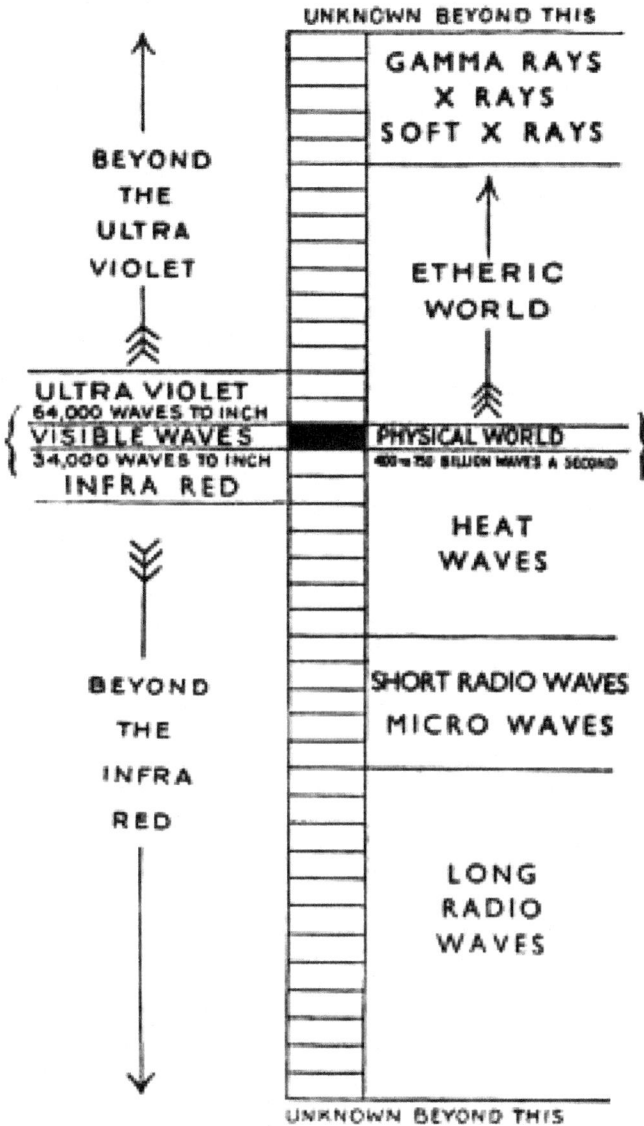

UNKNOWN BEYOND THIS

GAMMA RAYS
X RAYS
SOFT X RAYS

BEYOND
THE
ULTRA
VIOLET

ETHERIC
WORLD

ULTRA VIOLET
64,000 WAVES TO INCH
VISIBLE WAVES
34,000 WAVES TO INCH
INFRA RED

PHYSICAL WORLD
400 - 750 BILLION WAVES A SECOND

HEAT
WAVES

BEYOND
THE
INFRA
RED

SHORT RADIO WAVES
MICRO WAVES

LONG
RADIO
WAVES

UNKNOWN BEYOND THIS

FIGURE 12-1

the "Etheric World," takes on a "higher" material state, very much like ice becomes fluid water, when its "rate of vibration" (temperature) increases.

Similarly, consciousness (Spirit) exists in physical form that is not discernible from our paradigm of three dimensions plus time. Consequently, matter that appears non-discernible when perceived at our level of vibration apparently takes on the form of substance at different vibrations. Our natural thinking is that these Spirits would be higher vibrations, but whether they are "higher" or "lower" remains unclear. Since 1906 we have tried to find them in higher vibrations.

"The substance that forms the bodies of spirit-people, vibrating more than five octaves higher than the violet ray." [74] —Edward C. Randall, 1906

The frequency of violet is from 668-790 THz (terahertz) and the visible spectrum is from 430-790 THz. So, anything higher than violet is non-visible ultraviolet. Each octave doubles the frequency so that "five octaves higher than the violet ray" would be 25.280 PHz (petahertz), or 2.528×10^{16} Hz. This is just on the borderline between extreme ultraviolet (EUV) and soft X-rays.

"It is most difficult for the human mind to comprehend that anything which sight or sense does not disclose is material. The idea that what we call space is substantial and real and composed of matter, the same as those things that are visible, present a proposition difficult of acceptance." [75] —Edward C. Randall, 1906

"Matter, in its higher and more refined vibrations, becomes a fluid of infinite suppleness and elasticity by endless combinations of which all bodies are engendered. In its primordial essence; invisible, impalpable, imponderable; this fluid, through successive transitions, becomes ponderable, and capable of producing, by powerful condensation, those hard, opaque and weighty bodies which constitute the base of terrestrial matter. This state of cohesion (as terrestrial matter) is, however, transitory. Matter, rescinding the ladder of its transformations, can as readily be desecrated and returned to its primitive fluid state." [76] —Edward C. Randall, 1906

According to our current science, this "more refined vibration at higher frequency" could exist in one of two "places": either in the dark energy and dark matter, which is postulated by many physicists to fill the universe, or in the electromagnetic energy of the "plasma," which is also postulated by other physicists to fill the universe.

DOES CONSCIOUSNESS EXIST IN PLASMA ENERGY? On the other hand, there is another state of matter where consciousness could exist without requiring alternative dimensions. Plasma is the fourth state of matter, other than solid, liquid, and gas. First discovered in 1879 by Sir William

Crookes (a spiritualist) and called *radiant matter*, plasma is a form of matter that exists in electromagnetic energy and takes the following natural forms:

- *Lightning*
- *St. Elmo's fire*
- *Upper-atmospheric lightning (e.g., blue jets, blue starters, gigantic jets, ELVES*—Emission of Light and Very Low Frequency perturbations due to Electromagnetic Pulse Sources)
- *Sprites—mythical, fleeting, and playful creatures that appear in mythology and Shakespearean plays. The name was first suggested by Prof. David Sentman of the University of Alaska-Fairbanks in 1994, and it stuck.*
- *The ionosphere*
- *The plasmasphere*
- *The polar aurorae (Northern Lights)*

More familiar examples of plasma would be plasma in *neon signs* and *plasma display panel (PDP) televisions*. However, plasma TV technology reached its peak in 2008 and has since been replaced with LCD and OLED television models.

Plasma is loosely defined as an electrically neutral medium of unbound positive and negative particles (the overall charge of a plasma is roughly zero). Although the particles are unbound, they are not "free" in the sense of not experiencing forces. Also, when the charges move, they generate electrical currents with magnetic fields, and as a result, they are affected by each other's fields. Plasma is the most abundant form of ordinary matter in the universe, most of which is in the rarefied intergalactic regions, particularly the intracluster medium, and in stars, including the sun. The dynamics of plasmas interacting with external and self-generated magnetic fields are studied in the academic discipline of magneto-hydrodynamics (MHD).

DOES CONSCIOUSNESS EXIST IN DARK ENERGY? Since the 1990s, observations have shown that the universe is expanding at an accelerating rate, and so far *dark energy is the most accepted hypothesis to explain this*. This theory says that everything we can observe is made up of light energy, which according to mathematical calculations composes about 4.6 percent of the known universe. But the other 96 percent of the universe that we cannot yet discern is called *dark energy (72%)* and *dark matter (23%)*.

Now, although we can mathematically prove their existence, *we cannot yet discern dark energy or dark matter.* Consequently, we can't yet bring them into 3-D perspective of reality. But this undiscerned 96 percent of the universe, which is entirely outside the materialist paradigm, gives us plenty of room for both consciousness and the afterlife to exist in. Yet the existence of dark

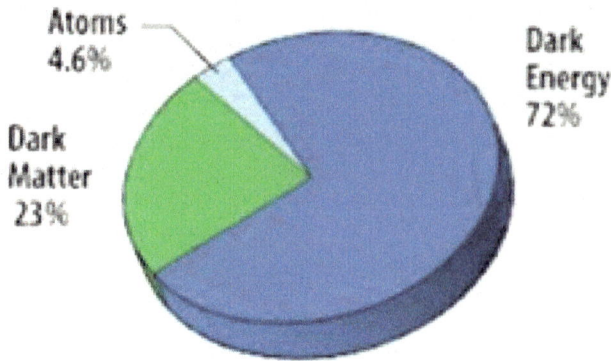

The ZERO-POINT FIELD

energy and dark matter signify that our understanding of physics is incomplete, and we will need a new idea as profound as *general relativity* to explain these mysteries. This is where the conscious universe/biocentrism theory make physics "complete." In fact, **string theory** and **M- theory** postulate that there are **eight additional unobserved dimensions out there** in this vast "undiscovered country."

THIRD AXIOM: *Classical Newtonian physics requires a consciousness that exists outside their limited perceptive box of 3-D plus time. Certainly that can easily exist within the 96 percent of the energy of the universe that is not yet discerned, the dark energy field, which pre-exists the evolution of the brain. That easily extrapolates to "consciousness came first," before brains; or it can exist in plasma energy, which pervades "empty" space.*

This brings us to the second postulate:

SECOND POSTULATE: EVERYTHING IS COMPOSED OF CON-SCIOUSNESS, NOT MATTER. *Consciousness is all there is... Consciousness is also non-physical, non-local, and of necessity, non-temporal. It exists entirely outside of time, space, and physical matter.*

THIRD POSTULATE: TIME IS ONLY A PERCEPTIVE ILLUSION.

For me this is a conundrum. I learned to navigate as a sailor during the Vietnam War. I used precise time to locate my position on the earth's surface using spherical geometry and carefully timed celestial sights. So for me, as a navigator, time is a very real entity. Yet quantum theory understands time to be an illusion that does not exist in any absolute sense.

There is no "absolute time" somewhere out there measuring the history of the universe, nor any reliable measure of when anything actually happens.

Extending this thought, it is obvious that there can be no actual beginning or ending to anything as measured by time. For example, **what was here a week before that big bang?** So, philosophically, from our current logical perspective, **time can have no beginning, or ending.**

*Einstein's special and general theories of relativity demolished the idea of time being a universal constant, in **1905–1915**. Consequently, the past, present, and future were no longer absolutes.*

Yet Einstein's theories seemed incompatible with quantum electrodynamics, which was fully developed eighty years ago in 1934. Thirty-six years later in 1970, physicist John Wheeler, at Princeton, and the late Bryce DeWitt, at the University of North Carolina, developed an extraordinary equation that provides a possible framework for unifying relativity and QED. Interestingly, when the Wheeler-DeWitt equation is applied, time just disappears from the mathematics. Maybe that is showing us that the fundamental description of the universe is simply timeless.

No one has yet succeeded in using the Wheeler-DeWitt equation to fully integrate QED with general relativity. Nevertheless, a sizable minority of physicists believe that any successful merger of relativity and QED will ultimately describe a universe in which there is no time. This "problem of time" gets deeper when you realize the laws of physics don't explain why time should always point to the future. All the laws of physics, in classical Newtonian physics, Einstein's physics of relativity, or QED, work equally well if time runs backward or forward.

Einstein's famous equation $E=MC^2$, which explains how energy multiplied by time "creates" matter, also shows that when you take "time" (C) out of the formula, there is no matter and only energy remains. Consequently, when energy is observed by "timing" it (measuring it in relation to time), this act of observation creates the matter. But when you stop the process of observation, the matter ceases and only waves of potential energy remain.

When your mind says, *"Wait a minute, I don't get that,"* the short answer is *"Just do the math."* Maybe this has to be taken on faith, but most physicists agree with the third postulate, that time is an illusion.

THIRD POSTULATE: TIME IS AN ILLUSION

FOURTH POSTULATE: EVERYTHING IS CONNECTED

QUANTUM ENTANGLEMENT, NON-LOCALITY, and NON-LOCAL SPACE: Erwin Schrödinger first spoke of non-locality in the 1920s when he described how things could be separate and non-separate at the same time. And since the 1930s three physicists had also observed a subtle kind of information transfer known as the EPR correlation, or simply "entanglement." This phenomenon was first discussed in their paper mentioned earlier, "Can

Quantum Mechanical Description of Physical Reality Be Considered Complete?" (1935). Einstein, Podolsky, and Rosen argued that missing elements of reality (*hidden variables*) must be added to equations of quantum mechanics to explain the entanglement.

ENTANGLEMENT: Here is a simple anecdotal explanation: Consider an unstable spin-less particle that decays into two particles, each of which then speeds off in opposite directions ("spin" is the measurement of the particles' angular momentum). The laws of conservation of energy require that the spins of the two particles must add up to zero (the spin of the original particle). This means that if one particle had an "up" spin, the other particle must have a "down" spin. The laws of QED require that in the absence of measurement, neither particle has a spin but remains in the quantum state until a measurement is made. When we measure (observe) the spin of either particle, then they both must leave the quantum state (wave) and become particles (matter), just as happens in the double slit experiment discussed earlier. This means that eventually, when the two particles are separated by several thousand light-years, when one particle is "observed" and it changes from a wave of potential into a particle of matter, the other particle MUST ALSO "COLLAPSE" FROM THE WAVE FUNCTION (precipitate into matter) AS A PARTICLE WITH SPIN IN THE OPPOSITE DIRECTION, INSTANTANEOUSLY. So, the particle we observed apparently communicates instantaneously (faster than the speed of light) across the entire universe. Einstein called this "spooky action at a distance."

In 1952, David Bohm added those "hidden variables," which EPR called for, to the Dirac equation. But moving along to 1964 and John Bell's theorem.

BELL'S THEOREM EXPLAINS ENTANGLEMENT: In 1964, John S. Bell, a physicist at CERN, in an effort to disprove entanglement published a mathematical "proof" for entanglement known as "*non-locality*." He challenged the world of physicists by saying in effect that, "If entanglement is true, then this must be true also." Beginning in 1972, other scientists have repeatedly proven this *non-locality* to be valid, and by 2009 this proof has been replicated nine separate times. John Bell would have received the Nobel Prize for this in 1990, but he died before the prize, which is not awarded posthumously, could be awarded. The theory was proven
- **1972** Freedman and Clauser,
- **1981–82** Alain Aspect,
- **1998** Tittle's group at Geneva
- **1998** (again), **2000, 2001, 2007, 2008, 2009**. So since 1998, these six replications have been made.

The replication of proof for Bell's theorem not only verifies the truth of *entanglement* and *non-locality*, which means that space itself is a perceptive illusion, but it also allows only one rational explanation, *that the universe itself is conscious (intelligent)*.

Since then, experiments with photons have repeatedly replicated the results of non-locality confirming these "entanglement" correlations do exist, and provided strong evidence for the validity of QED. As a result, particle physicists have now proven that those two "related" subatomic particles in the explanation of entanglement above, no matter how widely they are later separated, even by a distance of thousands of light-years, are not actually separate (i.e., they are *non-local;* while they appear within our perception to be dimensionally widely separated, they are in fact in the same locality). So, thinking long enough about that forces the realization that space itself is an illusion. An easier way to express this is that *the particles themselves don't have a specific location, but exist simultaneously in multiple localities, as a wave of probability.*

That's right, they appear to be in more than one place at the same time.

While the mathematics remain indecipherable to non-mathematicians, the implications of Bell's theorem profoundly affect our basic worldview and tell us clearly that there is no such thing as a separate part (the space between things or the separation is completely an illusion).

At the subatomic level everything is interconnected and information moves instantly across vast distances, and this is called quantum entanglement. This explains how psychic phenomena and mediumship work and how people have experiences of "direct knowing."

FOURTH POSTULATE: SPACE IS AN ILLUSION...EVERYTHING IS CONNECTED. *All parts of the universe are connected in an intimate and immediate way.*

Interestingly, this also verifies the theory received through the mediumship of Emily French in the 1890s and published in 1906, by Edward Randall, but originating from Dr. David Hossack, M.D., a surviving consciousness in the afterlife (founder of Columbia Medical School) in which Dr. Hossack said that *the entire universe was capable of intelligent thought.* At the time, in 1906, no living scientist here in the physical believed this. But Dr. Hossack in the afterlife was privy to things obscured from us while living in the physical. Today 110 years later, many living scientists do believe this. More profoundly, this clear statement of the coming science one hundred years early also verifies the Spiritualist concept that after-death communication with surviving consciousness through mediums is indeed valid. Which brings up our next topic.

WHAT HAPPENS TO INDIVIDUAL CONSCIOUSNESS AT DEATH? The energy of our consciousness cannot simply cease to exist. In strict accordance with the laws of classical Newtonian physics, it must continue. The Law of Conservation of Energy states that *energy never disappears; it only changes form, with no beginning or ending.*

Cardiac surgeon Dr. van Lommel, M.D., summarizes the current state of survival theory among scientists collating quantum mechanics and the NDE as:
> *"Most people still believe that death is the end of everything... That used to be my own belief. But after many years of critical research into the stories of NDErs, and after careful exploration of current knowledge about brain function, consciousness, and some basic principles of quantum physics, my views have undergone a complete transformation. I found the most significant finding to be the conclusion of one NDEr: 'Dead turned out to be not dead.' I now see the continuity of our consciousness after the death of our physical body as a very real possibility."* —Dr. Pim van Lommel, Cardiac Surgeon, *Consciousness Beyond Life*

FIFTH POSTULATE: THERE IS ONLY CONSCIOUSNESS

Since everything is composed of consciousness and not matter (Second Postulate) and since everything is also connected, or entangled, as described by non-locality (Fourth Postulate), then this Fifth Postulate derives that *there is only consciousness.* This in turn becomes a theoretical basis for the existence of both Universal Consciousness and the unity of all things... Call it a "singularity," if you will.

THIS IS NOT RELIGION: Now, please keep in mind this is not a religious book; this is a scientific book... So, when I speak of the concept of "Universal Consciousness," "The Force," or "God," I am merely restating a scientific fact of quantum electrodynamics (QED), a fact proven beyond a reasonable doubt. But I am not stating a religious belief... Instead, this is the same paradox Einstein, Erwin Schrödinger, Heisenberg, Bohr, Pauli, and Dirac struggled to explain, and what Feynman calls "the only mystery." Yet: *this consciousness, which underlies everything, also provides a locus for the afterlife survival of our individual consciousnesses. Apparently, this "afterlife" is located in alternative dimensions of dark energy, but can be accessed (communicated with) through mediumship. And that mediumship is accomplished using the same bi-phasic mechanisms of the will channel and the consciousness channel (the observer effect or Third Postulate) which provide the hidden variables precipitating the state vector collapse creating our reality.*

Actually, there is nothing new here, just well-proven scientific facts that remain "officially unrecognized" by the die-hard materialists, but prove consciousness and consciousness survival in an afterlife. Personally, after having actually been there in the afterlife during my NDE discussing all this with the *Being of Light,* I already know viscerally that everything in our universe is consciousness. And I also know, ***beyond any reasonable doubt,*** that consciousness survival in an afterlife is real.

Further, as a practicing medium, I daily communicate with discarnate consciousnesses still living in that same afterlife. Now, if you can accept each postulate on its own terms, you will make the same conclusion as Dr. Pim von Lommel that "DEAD TURNS OUT NOT TO BE DEAD"...or as I and the Spiritualists say,

"There is no death, there are no dead."

FIFTH POSTULATE: THERE IS ONLY CONSCIOUSNESS

Here is a video with Nobel Laureate Richard Feynman, who received the prize for his mathematical explanation of QED.

- **Richard Feynman Lecture on Quantum Electrodynamics: QED. 1/8** Uploaded on Jan 9, 2011
  Feynman gives us a lecture on quantum electrodynamics, the theory of photons and electron interactions which incorporates his unique view of the fundamental processes that create it. One of the three winners of the 1965 Nobel prize in physics for his work.
  Part 1: https://youtu.be/LPDP_8X5Hug,   Part 2: http://www.youtube.com/watch?v=rKjpk3

CHAPTER 12

SCIENCE OF BIOCENTRISM

*"The only thing we can perceive are our perceptions... In other words, **consciousness is the matrix upon which the cosmos is apprehended.** Color, sound, temperature, and the like exist only as perceptions in our head, not as absolute essences. In the broadest sense, we cannot be sure of an outside universe at all."* —George Berkeley (1685–1753)

*"All matter originates and exists only by virtue of a force which brings the particle of an atom to vibration and holds this most minute solar system of the atom together. We must assume behind this force the existence of a conscious and intelligent mind. **This mind is the matrix of all matter.** "*[77] — Max Planck (1858–1947), Father of Quantum Mechanics

"The universe and the observer exist as a pair. I cannot imagine a consistent theory of the universe that ignores consciousness."[78] — *Andrei Linde, Physicist, Stanford University*

WHAT IS BIOCENTRISM?

Biocentrism is a new concept, sometimes called the *Biocentric Universe Theory*, that has swept through biology and physics since 2007. It simply places life first. Instead of life (consciousness) arising from dead matter, as the *Big Bang Theory* postulates, and as materialist dogma dictates, this biocentric viewpoint postulates that consciousness came first and "created" everything else. Biocentrism reflects how the universe is designed to support life, and that this is no accident.

Logically, when we consider the materialist explanation that *consciousness accidentally arose from unconscious matter,* but for which there is no cogent explanation of how this could have happened, suddenly, such a ridiculous idea is seen for what it is, simply unbelievable foolishness and not science.

One might as well say, "Consciousness *magically* arose from unconscious matter." But magic is not science, no matter how much materialists may need magic to prop up their archaic theories.

"No one has yet been able to explain how consciousness could emerge from mindless matter." —Henry Stapp, Ph.D. Theoretical Physics Group, Lawrence Berkeley National Laboratory, University of California, Berkeley

Further, in order to suit the requirements of a scientific fact, as materialists like to pretend that materialism is, this harebrained idea needs to qualify as a scientific fact.

1. First, it must be a logical possibility...which it is not.
2. Second, there needs to be evidence for it... The only evidence materialists have is that consciousness exists at all, which is in no way determinant that it must then have arisen from dead matter.
3. Third, there also needs to be no evidence against it... But instead there is already a great deal of evidence against it.

Therefore, by our scientific requirement for "proof beyond a reasonable doubt," there is no way that this ridiculous hypothesis of conscious life arising from dead matter can be considered as a scientific fact; it is only so much materialist BS. Instead, we again find that *the only thing supporting this materialist dogma as being a fact are the unrecognized superstitions of the materialist's unrecognized religion.*

BIOCENTRISM: The foundations for this theory of biocentrism were first proposed by physicist John A. Wheeler, in following what Eugene Wigner calls the "orthodox interpretation of quantum electrodynamics," which is the Copenhagen Interpretation as formalized by the mathematician John von Neumann (1955 and 1932)[79], and as rationalized to agree with relativity by S. Tomonaga (1946)[80] and J. Schwinger (1951)[81].

Wheeler, the physicist who gave us the terms "black hole" and "worm hole" and worked on the Manhattan Project, suggested that reality is created by observers and that *"no phenomenon is a real phenomenon until it is an observed phenomenon."*

He coined the term "participatory anthropic principle" (PAP) from the Greek "anthropos," or human. He went further to suggest that:

"We are participants in bringing into being not only the near and here, but the far away and long ago."

This claim was considered rather outlandish until his thought experiment, known as the "delayed-choice experiment," was tested in a laboratory in 1984. This experiment is a variation on the famous double-slit experiment in which the dual nature of light is exposed (depending on how the experiment is measured and observed, the light behaves like a particle, a photon, or a wave), as explained under the First Postulate in Chapter 11.

But unlike the original double slit experiment in Wheeler's version, the method of detection is changed AFTER a photon had passed through the double slit. This experiment showed that the path of the photon was not fixed until the physicists made their measurements. In other words, the light wave/photon stayed in the quantum state even after passing through the slit, and remained in an indeterminate state **until it was finally observed by a living consciousness.** *This means that the observation was retroactive…moved backward through time. The results of this experiment, as well as another conducted in 2007, proved what Wheeler had always suspected:* **An observer's consciousness is required to bring the universe into existence.**

It also shows that the arrow of time is itself an illusion. All of this means that a prelife Earth would have existed in an undetermined state (wave function), and a prelife universe could only exist retroactively. Of course most of that is simply too much for a materialist, who sees it all as impossible. But unfortunately for their religious BS of materialism, the data says that it is so. Consequently, all they can do is ignore the data that disagrees with their chosen BS of materialism.

This idea took on real power when it was taken up by biomedical researcher Robert Lanza[82]. *Lanza has been on the frontier of cloning and stem cell studies for more than a decade, so with his cutting-edge stature, his book* Biocentrism *is generating controversy at a whole new level. Biocentrism holds that consciousness plays a central role in creating the cosmos. Dr. Lanza says,*

> **"By treating space and time as physical things, science picks a completely wrong starting point for understanding the world."**

As mentioned earlier, Dr. Lanza's theory of biocentrism has seven principles:

1. **What we perceive as reality is a process that involves our consciousness.** *An "external" reality, if it existed, would by definition have to exist in space. But this is meaningless, because space and time are not absolute realities but rather tools of the human and animal mind. John Bell's theory proves that there is no locality (no space), only non-locality.*

2. **Our external and internal perceptions are inextricably intertwined.** *They are different sides of the same coin and cannot be divorced from one another.*

3. **The behavior of subatomic particles, indeed all particles and objects, is inextricably linked to the presence of an observer.** *Without the presence of a conscious observer, they at best exist in an undetermined state of probability waves.*

4. **Without consciousness, "matter" dwells in an undetermined state of probability.** *Any universe that could have preceded consciousness only existed in a probability state.*

5. **The structure of the universe is explainable only through biocentrism.** *The universe is fine-tuned for life, which makes perfect sense as life creates the universe, not the other way around. The "universe" is simply the complete spatiotemporal logic of the self.*

6. **Time does not have a real existence outside of animal-sense perception.** *It is the process by which we perceive changes in the universe.*

7. **Space, like time, is not an object or a thing.** *Space is another form of our animal understanding and does not have an independent reality. We carry space and time around with us like turtles with shells. Thus, there is no absolute self-existing matrix in which physical events occur independent of life. Again this is proven by John Bell's non-locality.*

Obviously, understanding this biocentric universe theory requires a radical change in the way we view the world and our place in it, because it proposes that the physical universe evolves in tandem with the evolution of Earthly life. **It postulates that the universe exists specifically in relation to us.** *This has been compared to how the position and appearance of a rainbow is dependent upon the position of the person seeing it. The rainbow can only be seen from certain angles.*

Most of us have been taught the materialist viewpoint that the universe is a collection of particles "out there," and that these particles (atoms and molecules that have been around far longer than the Earth) came together billions of years ago to create the first life forms.

Biocentrism, on the other hand, considers these ideas of a reality "out there" to be entirely unfounded assumptions, not supported by any empirical evidence. This is consistent with the fact that materialism—based on those unsupported presuppositions of reality, locality, causality, continuity, *and* determinism, *all of which we showed in Chapter 3 were entirely falsified by QED—is itself an unfounded assumption.*

Biocentrism also explains how the universe could have an extremely simple beginning, while today appearing to be astonishingly complex as well as precisely "fine-tuned" for the existence of life. Also, it is completely consistent with everything we truly know in science, and solves other mysteries intuitively as well. There is also no conflict regarding the empirical findings of physics. The only conflict biocentrism has with mainstream science involves the interpretation of these empirical facts (the materialist BS).

Since the 1920s, quantum physics experiments have routinely shown that George Berkeley's argument, ***"Consciousness is the matrix upon which all else is apprehended,"*** *and Max Planck's statement that "Mind is the matrix of all matter,"* to be the correct perspective.

From the biocentric point of view, life, and particularly consciousness, creates the universe, and the universe could not exist without us (consciousness) in it.

175

As stated above, quantum electrodynamics is our most accurate model for describing the universe. And it also makes some of the most persuasive arguments that conscious perception is integral to the workings of the universe. Werner Heisenberg's famous uncertainty principle, which tells us that an unobserved small object (for instance, an electron or a photon—a particle of light) exists only in a blurry, unpredictable state, with no well-defined location or motion until the moment it is observed, and describes the phantom, not-yet-manifest condition as a wave function.

The wave function is Schrödinger's mathematical expression used to find the probability that a particle will appear in any given place. When such a property switches from possibility to reality, we say that the wave function has collapsed (the state vector collapsed). But what accomplishes this collapse? Merely a living consciousness observing it in any way. Experiments suggest that mere knowledge in the experimenter's mind is sufficient to collapse the wave function, converting possibility into reality.

Further, when particles are created as a pair, physicists call them entangled. And because of their entanglement, the particles share a wave function. When we measure one particle and thus collapse its wave function, the other particle's wave function instantaneously collapses also. Consequently, if one photon is observed to have a vertical polarization (its waves all moving in one plane), the act of observation causes the other to instantly go from being an indefinite probability wave to an actual photon with the opposite, horizontal polarity, even if the two photons have since moved great distances from each other.

BELL'S THEORY EXPLAINS ENTANGLEMENT: *As described above, the implications of Bell's theorem profoundly affect our basic worldview and tell us clearly that there is no such thing as a separate part (the space between things or the separation is completely an illusion).* **All parts of the universe are connected in an intimate and immediate way.**

The replication of proof for Bell's theorem also allows only one rational explanation, which is that the universe itself is conscious (intelligent). Lanza's view is fully in line with the perspective from QED that the observer plays a huge role in how reality is observed.

CONCLUSION:

Biocentrism answers the basic questions of:
1. What created the Big Bang?
2. What was here before the Big Bang?
3. What is the universe?

by saying, "The universe is an active, life-based process which does not exist outside of mind, and what came before the Big Bang assumes that time has an arrow of direction from past to future, but time itself is an illusion." Maybe we should think of time as a movie film already in the can. You can start at the beginning or the end and watch the reel, but the scenes already exist, past, present, and future at the same time.

176

So, then is biocentrism a scientific fact?
1. We have a logical possibility…
2. We have a great deal of evidence for it.
3. There is no evidence against it… It does not violate the principles of QED.

Therefore, by our scientific requirement for **proof beyond a reasonable doubt,** biocentrism is already an established scientific fact. And again we find that the only thing stopping recognition of biocentrism as a fact by the scientific establishment is their own fully unrecognized religious superstitions (materialist BS). Two short videos explain biocentrism at https://youtu.be/KL7-t9p8Zac

Is the opposing *materialist* idea that consciousness arose from unconscious matter a scientific fact?
1. It needs to be a logical possibility…which it is not.
2. It needs to have evidence for it… The only evidence materialists have is that consciousness exists at all, which is in no way determinant that it must then have arisen from dead matter.
3. It also needs to have no evidence against it… But there is a great deal of evidence against it.

Therefore, by our scientific requirement for *proof beyond a reasonable doubt,* there is no way that this ridiculous hypothesis of conscious life arising from dead matter can be considered as a scientific fact. As always, we find that *the only thing supporting this materialist dogma as being a fact is the Materialist's own fully unrecognized religious superstitions.*

As soon as the scientific establishment recognizes their *atheist* mechanistic *materialist* dogmas as the religion they represent, then scientists will move on. But as history has shown, this usually requires that they die first, which is why *science advances one funeral at a time.*

Here some videos discussing Dr. Lanza's biocentrism theories:
- **Dr. Robert Lanza discusses Biocentrism & Stem, YouTube** https://www.**youtube**.com/watch?v=4_LU0JG5I8M Jan 9, 2013—Uploaded. Biocentrism & Stem Cell Research
- **Dr. Robert Lanza on theory of Biocentrism - YouTube** https://www.**youtube**.com/watch?v=zI_F4nOKDSM Jun 7, 2011—Uploaded. This is part 1 of Robert **Lanza's** talk on Biocentrism at the Science and Nonduality
- **Dr. Robert Lanza at CCRI, 10/29/2013 - Part I - YouTube** https://www.**youtube**.com/watch?v=7z61aAmTpIk Dec 6, 2013—Uploaded, **Dr.** Robert **Lanza,** Chief Scientific Officer, Advanced Cell Technology.

- **Dr. Robert Lanza Part I (StemConn 2015) - YouTube**
  https://www.**youtube**.com/watch?v=W5I2xg_ct8Q
  Apr 28, 2015—Uploaded. Why is Robert **Lanza** not funded.
- **Biocentrism - Robert Lanza - YouTube** https://www.**youtube**.
  com/watch?v=YehIxgLNIJg
  Nov 12, 2014 - Uploaded
- **Dr. Robert Lanza, 3/24/13 - YouTube** https://www.**youtube**.
  com/watch?v=sf7NCVqh0EE
  Mar 24, 2013 - Uploaded

CHAPTER 13

WHAT ALL THIS MEANS

STAGGERING CONCLUSIONS:

One of the most fundamental scientific developments in the past three decades has been the repeated experimental confirmations of the principle of non-locality (John Bell's theorem). This, coupled with the implications of quantum electrodynamics (QED), presupposes a consciousness existing entirely outside the limited perceptive box of classical Newtonian physics (three dimensions plus time), and further postulates a conscious universe including consciousness survival in an afterlife.

Although dogmatic *Materialists* may continue to fearfully deny psi phenomena as being "IMPOSSIBLE" strictly because all psi violates their totally disproven First Principles of *Materialism*, this position of denial of psi is simply no longer tenable by any intelligent scientist. The six underlying presumptions of materialism (***reality, locality, causality, continuity, determinism,*** and ***certainty***) were all falsified fully eighty years ago by the modern science of QED: thus, *Materialism* can no longer be considered to be in any way a scientific perspective, yet can be considered as a religious perspective.

Instead, QED is now the most proven scientific theory of all time, and it supports both psi phenomena and consciousness as being the underlying matrix of all reality.

This single breakthrough of proving *non-locality* shows, unequivocally, that psi phenomena and consciousness survival are not in conflict with the established laws of science. Finally, we can say,

> **Modern science has proven consciousness survival (the afterlife) to be a fact.**

Consciousness survival is not at all as religions have portrayed it to be for thousands of years, as being a place of punishment and reward... Instead, survival in alternative dimensions is just the next stage of consciousness. This scientific truth will simply replace all superstition and religion regarding the afterlife. Unfortunately, churches will no longer have anything to sell (jump through our set of hoops and you can have eternity). This is obviously why last year (in the fall of 2015) the Fundamentalist Christian churches in America decided to no longer allow *near-death experience* books in their "Christian" bookstores. Obviously, this is what Dr. Schwartz would call ostrich vision…or let's hide from the truth.

*We still lack a reliable database with sufficient repeatable psi effects upon which complete theories might yet be constructed and refined. And while we also lack a comprehensive theory of consciousness itself, upon which any theory of psi must, inevitably, be built…on the other hand, we have, as Max Planck predicted, shown that the **underlying matrix of all reality** is in fact consciousness, and this consciousness is proven from our modern science to be just as real, and measurable, as magnetism and electricity.*

Cumulatively, everything we each believe that we see is the personal illusion we are mentally projecting as our universe. This reality we each sketch is entirely constructed of thoughts, and these are actually shared thoughts (collective consciousness) so that, like schools of fish, which suddenly turn together, the habitual perceptive fields of our surrounding culture causes those around us, through habit, to project a similar reality. As a result, we behold similar illusions (shared reality).

It is only when we lift the veil and look behind it (as I was privileged to do during my NDE, and as I continue to do each time I work as an evidential medium in after-death communications) that we can see the true reality, which in my opinion can most easily be stated as:
We are our thoughts, and our thoughts are the universe.

How this **underlying matrix of all reality** is structured was recently described by a leading cosmologist, former Astronaut Edgar Mitchell, and founder of the Institute for Noetic Sciences (noetic.org). Here is an edited paraphrase of his description of the universe as the *Quantum Hologram:*

> **THE NON-LOCAL QUANTUM HOLOGRAM:** *The underlying scientific case for mind-to-mind and mind-to-matter communications has been impressively well documented by several rigorous studies of these phenomena, i.e., Princeton Engineering Anomalies Research (PEAR), Stanford Research Institute (SRI), University of Virginia, and University of Arizona. These ongoing studies conducted over many decades provide compelling results showing staggering probabilities against chance.*

> *Consequently, this non-local quantum hologram, which is based on sound theory and recently verified through the practical application of fMRI, provides sufficient rigorous proof to postulate the theory that the quantum hologram is, in fact, a macro-scale, non-local, information structure (Akashic Record) that extends quantum electrodynamics to encompass all*

physical objects, including DNA molecules, organic cells, organs, brains, and bodies.

Further, the storage and transmission of non-local correlations and non-local quantum information can now be seen to be ubiquitous throughout the universe...such that the quantum hologram can properly be labeled as "nature's mind" and that the intuitive function we label in humans as the "sixth sense" or "second sight" (mediumship) which reads this record should properly be called the "first sense."

> **Any complex evolved organism which can form an intent (idea or wish) can, in fact, produce and often does produce non-local causal effects associated with that specific intent.**

It has been found that attention alone produces coherence in nature (resonant relationship, P-car, quantum coherence) that in some measure reduces randomness.

Finally, it can be postulated that non-locality in the universe is the antecedent attribute of energy and matter, which both allows perception and is also the root of the consciousness, which manifests in the evolved organisms existing in three-dimensional reality.

EMISSION AND REABSORPTION BY DARK ENERGY FIELD: *While investigating improvements to functional magnetic resonance imaging (fMRI), it was discovered that the phenomenon of emission/reabsorption of energy by all macro-level physical objects carries information about the event history of that object in phase relationships of the underlying interference patterns of the emitted quanta. The resulting mathematical formalism (theoretical model) is the same formalism used in holography, hence has been named quantum holography (QH). QH also provides a basis for explaining how the whole of creation learns, self-corrects, and evolves as a self-organizing, interconnected holistic system. Non-local quantum correlations observed in particle interactions exhibit the capacity to admit and absorb information at all scale sizes, guiding evolutionary processes with built-in storage and retrieval mechanisms for vast quantities of information. The phenomena of QH take place at all temperatures.*

The implications of QH are immense. QH provides a means for how dark energy and dark matter (now thought to be primordial energy and matter) arise out of the zero-point field. It suggests that it is the basis from which self-organizing systems learn, including life. The non-local quantum correlations observed in particles, and the non-local QH associated with molecular and larger scale objects, appear to serve the purpose of providing information at all scale sizes to guide evolutionary processes.

EVOLUTIONARY FEEDBACK LOOP: *The discovery of the quantum hologram as a solution, which seems to resolve so many phenomena, points to the logical conclusion that classical theory is incomplete unless it includes the subtle non-local components so deeply involved. This suggests that a major paradigm shift is now forthcoming in our twenty-first-century scientific ontology.*

Papers published by Marcer and Schempp propose a learning model both for DNA and prokaryote cells using quantum holography, and further suggest that evolution in general is driven by a learning feedback loop within the environment, rather than progressing through random mutations as postulated in Darwinian theories. This same solution to biological evolution was originally proposed by Lamarck in 1809, but was discarded by nineteenth-century materialist scientists in favor of the clockwork mechanistic solution of random mutations suggested by Darwin. —Paraphrase of former astronaut, Dr. Edgar Mitchell

Here are some videos with Dr. Mitchell:

- **Consciousness and the Quantum Hologram :** Dr. Edgar D Mitchell:
  Dr. Mitchell discusses the Quantum Hologram on this video.
  https://youtu.be/dAHXwPGTaIl
- **Edgar Mitchell - The Quantum Hologram Pt 1/3** Uploaded on Apr 11, 2009   .
  The late Dr. Edgar Dean Mitchell, D.Sc. (born September 17, 1930, died February 4, 2016,) was an American pilot and astronaut. As the lunar module pilot of Apollo 14, he spent nine hours working on the lunar surface in the Fra Mauro Highlands region, making him the sixth person to walk on the moon. Mitchell's interests include consciousness and paranormal phenomena. During the Apollo 14 flight he conducted private ESP experiments with his friends on Earth. In early 1973, he founded the nonprofit Institute of Noetic Sciences (IONS) to conduct and sponsor research into areas that mainstream science has found unproductive, including consciousness research and psychic events. https://youtu.be/-4U2sNtJKEU

WHAT THIS MEANS:

Literally, all the information that ever was is actually stored throughout the universe and can be accessed by those who have psychic abilities, or a sixth sense (i.e., you and I can tune into the P-car resonance signals). Also, each of us materializes in the physical instantaneously and continually in a holographic energy projection from the dark energy into this light energy. The following quotation is from the founder of quantum electrodynamics, Erwin Schrödinger, (parenthetical inclusions are my own interpretations):

> *"Inconceivable as it seems to ordinary reason, you—and all other conscious beings, as such, are the all in all."* (*The soul of the conscious universe*)

> *"Hence this life of yours which you are living is not merely a piece of the entire existence, but is in a certain sense the whole... Thus, you can throw yourself flat on the ground, stretched out upon Mother Earth, with the certain conviction that you are one with her and she with you. You are as firmly established, as invulnerable as she, indeed a thousand times firmer and more invulnerable. As surely as she will engulf you tomorrow, so surely will she bring you forth anew to new striving and suffering. And not merely 'someday.' Now, today, every day she is bringing you forth, not once but thousands upon thousands of times, just as every day she engulfs you a thousand times over..."*

The zero-point field (ZPF) continually generates the specific energy that perpetuates the illusion of the matter that is you...

> *"You move your hand and the universe carries out the subtle energy transformations from dark energy to light energy that constantly support the holographic illusion of matter, which is you here in space-time, and continues to instantaneously recreate the hologram of your hand in the moving positions and gives you the illusion of movement...you think and the universe performs the particle physics 'magic' to make it so."*[83] —Erwin Schrödinger, Founder of Quantum Mechanics

The above quotation is not from a medieval Christian mystic, but from one of the greatest scientific minds of the twentieth century, from the very mind that created quantum electrodynamics. Schrödinger realizes that each of us is part of the whole. Each of us is an integral part of the conscious universe and inseparable from it.

Here is another quotation showing how another great scientific thinker sees the same phenomenon:

> *"Up to now we have been looking at matter as such, that is to say according to its qualities and in any given volume—as though it were permissible for us to break off a fragment and study this sample apart from the rest...*

> *"It is time to point out that this procedure is merely an intellectual dodge. Considered in its physical concrete reality...the*

universe cannot divide itself but, as a kind of gigantic 'atom' it forms in its totality...the only real indivisible... The farther and more deeply we penetrate into matter, by means of increasingly powerful methods, the more we are confounded by the interdependence of its parts. Each element of the cosmos is positively woven from all the others...

"It is impossible to cut into this network, to isolate a portion without it becoming frayed and unraveled at all its edges. All around us, as far as the eye can see, the universe holds together, and only one way of considering it is really possible, that is to take it as a whole in one piece." [84] —Pierre Teilhard de Chardin

And yes, this time the quote is from a Christian mystic, but one who is also a world-renowned twentieth-century paleontologist. Yet both quotations, one from a physicist and one from a Christian mystic paleontologist, are striking at exactly the same chord, and that is: *We (each of us) are not separate from the whole.*

This is a philosophical application of *non-locality*, which shows that we as individuals are not separate from the Conscious Universe.

CONCLUSION:

Consciousness is the foundation of our universe, the underlying matrix upon which everything else is structured, just as Dr. Max Planck said in 1900.

- This Universal Consciousness is one inseparable whole (God, if you will, or "The Force"), yet minds and souls do individuate within that whole.
- Additionally, mind and soul can move from one body to the next (as experienced daily through transplant surgery), which shows that individual consciousness can transmigrate (reincarnate) from one physical flesh into another, just as I reincarnated back into the same body after my NDE.
- Therefore, the consciousness that transmigrated from one flesh to another must exist entirely separate from any physical flesh, and merely occupies that flesh.

Consequently, consciousness is primary, and physical flesh (physical reality) is secondary. And from these conclusions, consciousness survival in an afterlife is simply automatic.

There is no death, there are no dead.

Further, if the entire universe is also one conscious whole, as inferred by QED and *non-locality*, and also by nearly all mystical traditions, then the individual conscious mind must actually exist totally outside the body, and we are only *perceiving* it to be inside the body. Hence, we are each an eternal consciousness (spirit) currently having a physical experience here on earth.

"...the existence of a hidden field (implicate order) of non-physical consciousness, occupying as yet undiscerned additional dimensions, which are outside the visible reality (explicate order) defined by 3-D plus time, has been proven scientifically by the recent repeated replication of non-locality, the studies of the near-death experience, and after-death communications as demonstrated by triple-blind laboratory experiments testing evidential mediumship. This hidden field (implicate order) of non-physical consciousness, also provides the matrix upon which the explicate order of observed reality is continually manifested." —The author, Alan Hugenot

So, there is now an emerging worldview that agrees with Max Planck that consciousness is the underlying matrix of reality, and this matrix is composed of a multilevel vibrational field (or fields) that is (are) *alive, conscious, and intelligent.* And this *infinite intelligence* is there for each of us individually to tap into.

Our challenge, as honest scientists, is to discover, through careful science, how we can all interface with this matrix of consciousness.

Honest *Materialist* scientists, realizing that their previous mind-set of refusing to look at conflicting data *a priori* was not at all science, but was a shallow religious belief (like the Pope refusing to look through Galileo's telescope), should instead now simply adopt the late Dr. Carl Sagan's healthy skepticism of *"we don't yet know."*

Instead of fighting the presumed lunatic fringe of parapsychological science, intelligent scientists now realize that the die-hard *Materialists* are merely trying to prop up a dead religious theory, and one which has been disproven by QED nearly a hundred years ago. Intelligent scientists now also see that this *Materialist* religion also pushes its believers over the edge into the irrational lunatic fringe..

Understanding that *Materialism* is a religion rather than science, explains the irrational behavior of *Materialist* believers and their zealous desire to "kill the Infidel."

Intelligent scientists should now, instead, step back from such irrational people and divorce themselves from such foolish, childish, and abnormal behavior.

Finally, intelligent scientists should begin viewing psi phenomena as belonging to an aspect of reality about which we know very little, but which has proven to be real (despite any hopes to the contrary). This is the only honest way for a true scientist to work. Such a humble change of perspective away from a fear-driven, pretended certainty of *Materialism*, based on absolutely no data, over to a healthy scientific skepticism open to new discoveries will allow open examination of the existing scientific data developed by decades of rigorous parapsychological scientific research.

From such honest, open-minded inquiry and patiently following where the data leads, rather than rejecting it in advance because *"we already know it is impossible"* will bring such honest scientists to new discoveries of truth and new laws of physics beyond the Newtonian paradigm, which will then raise their stature in the scientific community and change the world for all time.

Why would any intelligent scientist miss the gold rush on the frontier of knowledge, in discovering the 96 percent we now realize that we don't yet know…just to uphold the archaic dead religion of Materialism, merely because that is what they were taught by their university mentors?

How stupid is that? To miss out entirely on the intellectual property shortly to be discovered in the conscious universe gold rush…in order not to be ridiculed and laughed at by the *Materialist* losers and pseudoscientist skeptics?

Well, personally I am not so fearful nor intimidated by *Materialist* BS. I could not care in the least what *Materialists* wish to believe. I am retired and don't need tenure, so I don't care what some calcified *Materialist* senior professors hope to know instead of the truth…or what the outspoken pseudoscientist skeptics want to say about me.

THE LUNATIC FRINGE:

But just who are these pseudoscientist skeptics, these self-appointed "protectors of *Materialist* science"? Frankly, they are a close-knit group of friends, but the most motley crew of jackasses I've ever investigated, and to be politically correct, I will admit that one of them is not a jackass…she is a Jenny Ass.

One example, of the last few years, is a magician named Penn Jillette, not a scientist at all or even a college graduate (although he does hold a degree from Ringling Brothers, Barnum & Bailey Clown College). But while taping a television interview (where Penn was supposed to be the surprise guest skeptic), this graduate clown told Dr. Gary Schwartz, who holds a legitimate Ph.D. from Harvard University and was a professor at Yale, that he (Penn) had checked Dr. Schwartz out and that he (Penn), in his expertise as a *Materialist* skeptic, had found that Dr. Schwartz did not graduate from Harvard… This is pure, unmitigated bullshit. But that is what magicians are always selling to the crowd.

Now, technically, this otherwise bald-faced lie is true. Dr. Schwartz did originally graduate from Cornell University (after studying under Dr. Carl Sagan, whom Materialists idolize), where he received his bachelor's degree. Later, Dr. Schwartz took his postgraduate doctorate degree at Harvard... So, technically (and only in a technical sense), Dr. Schwartz did not do his undergraduate work at Harvard and so did not "graduate" from Harvard. But, here we have a carnival clown magician, who never graduated from any real college (other than Ringling Brothers, Barnum & Bailey's Clown College) telling a legitimate Harvard Ph.D. in a taped interview that his doctorate was a lie...

This entirely contrived innuendo that Dr. Schwartz lied about his doctorate shows the low level of scientific method used by the pseudoscientist skeptics and also the low level of truth provided by television, which is only interested in ratings. When this kind of bullshit is carefully staged just before a commercial break, it makes great television for the gullible audiences magicians are used to. Illusionists (magicians) think the average person is a fool. Luckily, the American public with any intelligence is moving away from the dumbed-down network TV, so perpetrating this level of stupidity will soon be lost to the skeptics.

But for Penn, a pseudoscientist *Materialist* and a graduate clown who makes his living as a "magician" (i.e., fooling people), to fool them again by perpetrating such a lying innuendo on television, although beyond crass as a human being, is merely his normal stock in trade. It shows the desperate levels of untruth to which *Materialist* believers will stoop merely trying to bring down the truth of honest scientists and thus preserve a façade of truth for their religious belief in *Materialism*.

When skeptics need to lie on television and in the media about serious parapsychological scientists like Dr. Schwartz, it shows the extreme poverty of their science. They are not merely losing the battle...it was lost eighty years ago with the advent of QED. Instead, they desperately hope that pure *bullshit* will continue to fool the American public. Fortunately, the level of intelligence has been rising in the general population, and these carnival illusionists now can only perform in Las Vegas, where the majority of the crowd are fools who already came there to lose money.

On the other hand, when one considers the intellectual level of television viewers who are willing to believe that the "survivors" (on network television) are really kicking one of their group off the island while the group continues to desperately forage for food, directly in front of a fully fed camera crew, one understands that this professional entertainer (Penn) may know just how stupid his audience really is. But Penn is really honest; the title of his personal TV show was *Penn and Teller...Bullshit*, admitting in the title of his show that his tripe is pure bullshit.

One of my student mediums told me yesterday that this same Mr. Penn is still working the "magic" stage in Las Vegas (in December 2015) and that throughout his performance, to which she innocently bought a ticket (not realizing he was a raving pseudoscience *Materialist* skeptic), he was still trying to tell anyone who would listen, including his captive Las Vegas audience who innocently bought a ticket hoping to be entertained, that **"all mediums are liars and fakes,"** not realizing perhaps that in the modern world, many of his audience might now be mediums. She said he is entitled to his opinion, but to preach it from the stage on Christmas Eve... He should be offering refunds to the audience for not delivering what they paid for.

Another arch-skeptic is James Randi (who has appeared as a guest on *Penn and Teller Bullshit*). He made a name for himself as a magician, "The Amazing Randi." He is also a lifelong friend of Penn. The rest (Dr. James Alcock, Dr. Michael Shermer, Dr. Gordon Stein, and Dr. Susan Blackmore) are not scientists either. None of these people (who do hold Ph.D.'s) has ever worked in a consciousness research lab after finishing their Ph.D. And although one or two might have hard science undergraduate bachelor's, they all changed their major to psychology before they graduated. None holds a doctorate in a hard science, and none has investigated the actual data developed by consciousness research in the last fifteen years. They are all professors, but not scientists. Fortunately, Gordon Stein is deceased, and all the other jackasses are in their sixties or seventies and are not going to change their tune now. But most younger *Materialist* skeptics, realizing we don't know 96 percent of what there is to know, are a good deal more cautious.

All I can say is, *"Materialist fools, get a clue."* I know all these pseudoscientists, all approaching seventy, will be vehemently decrying everything I have said, saying I am a pseudoscientist, assassinating my character, and vehemently vilifying me...just like they always do. But I (and the honest and intelligent frontier scientists of the world) won't be listening. I've personally been there to the afterlife and so I KNOW, unequivocally, that this alleged "impossibility" is actually real.

RESOURCES:

Here are the URL's fpr three videos available on YouTube where I further explain all this, and three additional videos which discuss the ongoing research at the Institute of Noetic Sciences and the Laboratory for Advances in Consciousness and Health, Department of Psychology, University of Arizona:

"ON THE NATURE OF CONSCIOUSNESS & SURVIVAL" Interview with Dr. Alan Ross Hugenot (October 2015) 55 minutes. https://www.youtube.com/watch?v=yByEQfaD314

"BEYOND OUR SIGHT" *Dr. Alan Ross Hugenot discusses the near-death experience and mediumship, along with Dr. Dean Radin and four other mediums and NDE survivors (shown at IANDS 2014, 57 minutes)* https://www.youtube.com/watch?v=xpSuO8DtiMM

"SCIENCE OF THE AFTERLIFE" *Dr. Alan Hugenot discusses his book* THE DEATH EXPERIENCE: What It Is Like When You Die. http://www.youtube.com/watch?v=sG8RAVh4VwE

Dr. Gary Schwartz, Director of the Laboratory for Advances in Consciousness and Health, Department of Psychology, University of Arizona, Discusses the Mystery of Consciousness https://youtu.be/b1OeVvZCKVQ

Dr. Dean Radin discusses the work we have been doing together in the Consciousness Research Lab (at noetic.org) VIDEO CLIP (2 minutes) at http://vimeo.com/113981492

Dr. Arnaud Delorme *discusses the MEDIUMSHIP SCIENCE we have been working on together at Noetic.org VIDEO CLIP (2 minutes)* http://vimeo.com/113345358

APPENDIX A

This excerpt from my prior book, *THE DEATH EXPERIENCE: What It Is Like When You Die,* explains how the history of religious thought has explained the near-death experience. And further how that explanation has become the basis of all the world's religions.

THE WEIGHING OF THE HEART "La pesée du coeur"

ANCIENT EGYPTIAN WEIGHING OF THE HEART:
On my desk there is a print of an ancient Egyptian papyrus entitled *la Pesée du Coeur (shown on the following page),* the original of which is held in Paris at the Louvre. The papyrus illustrates the ancient Egyptian myth of "the weighing of the heart," the philosophy of which has been documented in surviving written artifacts from as far back as 4,500 years ago (2500 BCE)[85]. Considering that remnants of this same ancient myth also appear even earlier in the *Bhagavad Gita,* dated by scholars around 3000 BCE (and archeo-astrologically dated to as early as 5561 BCE), the myth portrayed in the papyrus is at least 5,000 years old and may be vastly older. Obviously, this myth is a basic part of mankind's inner psyche as understood by Carl Jung and Joseph Campbell.

This ancient myth also contains all the significant beliefs of most modern religions (Christian, Muslim, Jewish, Hindu, and Buddhist). Yet the myth is also obviously describing **the same** *life review* **as reported by the twenty-first-century NDE**, which examines how the recently deceased feel about all the events that transpired in their immediately prior life. Here is that papyrus with an explanation following that describes the weighing of the heart.

THE LIFE REVIEW:

One of the most intriguing things to come out of the recent studies of the near-death experience (**NDE**) is the persistent occurrence of the *life review*. This episode, which the ancient Egyptian religion named the *weighing of the heart*, occurs in the accounts of hundreds of NDEs, and provides the participant with a review of all the significant events of their lifetime. Although it is not recalled by all NDE survivors, when it does occur, it takes place shortly after the consciousness has left the physical body. After conclusion of this life review, the NDE continues with an important question:

> **Was it enough?** or **Have you learned enough that you are ready to move on?**

If the answer is "No," and the deceased individual still has unfinished business in the prior life, then the decision is made to return to that prior life and reincarnate *in the same body*. Apparently, if the answer to this question is "Yes," then the person moves on into the afterlife, because no one who answered "Yes" has returned to the prior life to report an NDE.

Ancient accounts of this life review occurring in the NDE have been discovered throughout the history of written literature and include several NDEs recorded by highly revered early Christian fathers. It is evident that earlier civilizations knew of NDEs, including the early Christians in the three centuries before the Roman emperors began to impose the Dark Ages, illiteracy, and "group think" onto Western Christianity after 390 AD, with Emperor Justinian finally outlawing belief in NDEs and transmigration (reincarnation) after 553 AD. But these ancient NDE accounts clearly agree with the modern NDE of the twenty-first century, all telling the same story of what lies just beyond death.

ALL AFTERLIFE SALVATION MYTHS BEGAN HERE:

This papyrus is clearly describing the life review portion of the near-death experience. So, it is then obvious that the surviving parts of the ancient Egyptian legend depicted in the papyrus which still appear in the salvation myths of each of the world's religions came from prior knowledge of the NDE.

> *"...man is destined to die once, and after that to face judgment."*
> Hebrews 9:27 (NIV)

La Pesée du Coeur
Photo by Author

THE WEIGHING OF THE HEART: This papyrus shows the goddess Maat, representing the divine attributes of truth, honesty, justice, and fairness, observing a balance-scale that is weighing the heart of a disincarnate individual just recently deceased. On the other side of the balance, being weighed against the heart, is Maat's feather of righteousness. Maat holds in her left hand the key of life (ankh), ready to award it to this deceased person if she finds the weight of their heart to be "light as a feather."

(She is asking him whether his experiences in the physical life have given him enough trials, tribulations, and lessons to achieve the growth necessary for his spirit to finally be at peace so that his heart is therefore light.)

Standing on a pedestal on the side of the deceased, and shown as a monkey, is Horus, the son of the god Osiris. Horus is himself a god who was born of a virgin (immaculate conception). His mother, Isis, was impregnated by the spirit (Holy Ghost) of the deceased god Osiris. Horus is actually waiting to tilt the scale in favor of the deceased. This *Son of God* is ready to literally "make up for the sins of the deceased."

TWO POSSIBLE OUTCOMES:
If the deceased's heart is not weighted down with the burden of sin, then Maat will give him the key of life and the deceased will travel on into the heavens to the stars Sirius (Isis) and Orion (Osiris) to rejoin the gods from which he came, and live with them in the heavens eternally, having after many lifetimes escaped the wheel of repeated lives and finally achieved Buddhahood.

On the other hand, if his heart is still weighted down with the burdens of this life, then his experiences in the physical life have not brought him enough pain to garner the gnosis necessary to no longer care about the things of this physical existence in time and space. And therefore, because his heart is still heavy, he will have to travel back through the underworld led by the guide dog Anuket to be "born again" and reincarnate as a human for another physical lifetime of seeking wisdom from the trials of physical life.

CHRISTIAN "LAST JUDGMENT": The Christians believe that after death there is a final judgment where Christ, who like Horus in the papyrus is believed to be the son of God and born of a virgin, makes up for one's sins. At the Christian *Last Judgment*, if you are found righteous you are given the "crown of life" (the ankh), and you then go to be with God (Isis and Osiris) in the heavens (the stars Sirius and Orion). But if your sins are too great (your heart is too heavy) and you don't seek help from Christ (or Horus) to tip the balance in your favor (propitiation), then you are condemned to go down into the underworld (Hades or hell). The only part of the Egyptian myth missing from today's Christian version of the salvation myth is the opportunity for rebirth into another physical life (reincarnation). Otherwise, the twenty-first-century Christian salvation myth is identical to the ancient Egyptian afterlife myth, but with the addition of retribution, hellfire, and damnation in the underworld, which has been transformed into a place of eternal retribution.

Twentieth-century biblical scholarship on early Christianity has shown that reincarnation was the accepted doctrine in early Christianity. It is now well documented that reincarnation was taught by the early Christian fathers, including Origen of Alexandria (185–254 AD), who is considered one of the greatest of the early Christian theologians.

NOTE: This means that the theology of the early Christian church of the first century, which included reincarnation in their original Christian salvation myth, was IDENTICAL to the ancient Egyptian salvation myth from 2,450 years earlier.

Further, the Christian religion of biblical times was a reawakening of ancient enlightenment of Thoth-Hermes in the Hebrew community much as it was already being practiced in the Hellenistic Mystery religions throughout the known world for at least 500 years before Christ[86].

As the early church Christian Bishop St. Augustine (November 13, 354 AD—August 28, 430 AD) in 427 AD stated, three years before his death, this so-called "new" Christian religion was merely a rebranding of the pre-existing "pagan" religion:

> *"That which is called the 'Christian' religion existed among the ancients, and never did not exist, from the beginning of the human race until Christ came in the flesh, at which time the true religion WHICH ALREADY EXISTED began to be called Christianity."*[87] St. Augustine, *Retractions*, Taken from P.54 *The Fathers of the Church, Saint Augustine, The Retractions.* Brogan. Catholic University of America Press

Augustine's *Retractions* were written as a clarification in 428 AD, after publication of his seminal work, *The City of God* in 427 AD.[2] My copy of this statement is authenticated by the Roman Catholic Church that this is precisely what St. Augustine said (see the endnote).

Given the above documented facts, honest science causes one to ask, *How did the original salvation myth get so grievously altered and why?* Here is that history:

THE SINGLE-LIFE THEORY: Unfortunately, politics has always been the force behind any form of "organized" religion. One hundred and twenty-five years after St. Augustine, the Christian salvation myth was drastically reduced by the Roman emperor in order to allow only a single physical life. In 553 AD, at the Fifth Ecumenical Council (2nd Council at Constantinople), which was called by Emperor Justinian and not the sitting Pope, the doctrines of Origen regarding reincarnation were ostracized.[88] The Pope even refused to attend this council, and it was Justinian who pushed for this rejection of reincarnation. Yet the adoption of these "anathemas against Origen," which shunned all those who believed in reincarnation, were never recorded in the official minutes of this Ecumenical Council. So it appears that the attending bishops did not want to grant the emperor's desires.

However, Justinian claimed that these anathemas were in fact adopted at the council, and then set about enforcing this change in doctrine and belief throughout the Roman Empire. Anyone who disagreed was simply killed. Next, where Jesus had clearly spoken in the canon gospels about multiple lifetimes, the church now had to reinterpret his remarks. Instead of multiple lifetimes to work on your karma and get it right with God, a Christian believer now had to be spiritually "born again" in the same physical body during that single lifetime. But now add to this Original Sin.

ORIGINAL SIN AND OFFICE OF THE KEYS: Earlier, in the late fourth and early fifth centuries, Roman Catholics added the new idea of guilt called *Original Sin* and along with it the idea of retribution or *eternal damnation* (hell) as a punishment for sinners. There are no scriptural or historic records for the existence of these beliefs in the four hundred years of Christianity prior to that time. Yet now, taking together these two ideas, only one life and retribution for sin in the afterlife, gave the Roman emperors through the church great power to successfully control the populace through religion. The Roman Church also claimed the power of the *Office of the Keys*, saying that God had given the Roman Church the power of retaining a person's sins against him, even though he may have repented. The Church could now, by simply refusing to allow forgiveness, deny a person eternal life in heaven and force them to suffer the eternal damnation of burning in hell.

MUSLIMS ALSO BELIEVE IN A LAST JUDGMENT: The Muslim beliefs about death and the afterlife, formulated by Mohammad in the seventh century, a hundred years after Emperor Justinian had literally killed off all beliefs in reincarnation, naturally retained the same *single-life theory*, but unlike the Christians, with no savior (Horus/Christ) and no rebirth in this life. Instead, Muslims believe you must do all atonement yourself in one lifetime by performing the five pillars of the Islamic faith, and in the Day of Judgment (*weighing of the heart*) if your good deeds outweigh bad deeds, then you go to live with God in paradise.

> NOTE: Within Islam there are two small Shiite sects that still believe in reincarnation or takamous.[89] These are the Druse of Lebanon and Syria and the Alevis of Turkey, similar to Sufi Muslims. All are regarded as not being truly Muslim by other sects of Islam. Consequently, they have been continually harried and killed by both the Shiite and Sunni sects for over 1,300 years.

HINDU AND BUDDHIST SALVATION MYTHS LACK A LAST JUDGMENT: In the East, both Hindu and Buddhist salvation myths retain the numerous lifetimes of reincarnation. The Hindus retain a form of the *Weighing of the Heart*, believing that after a person dies, a supernatural being, Brahman (Maat), weighs the good and evil done by that person and assigns them their next place of reincarnation. Brahman can mitigate the negative effect of early mistakes in life if later in life many good deeds were done. In Buddhist thinking, which is an offshoot of Hindu beliefs, there is no actual *Weighing of the Heart,* but the soul cannot escape the "wheel of repeated lifetimes" until it is worthy of Buddha-hood, and this attained level of purification amounts to the same result as the *Weighing of the Heart.*

COMPARISON OF RELIGIOUS SALVATION MYTH BELIEFS:

ORIGINAL EGYPTIAN RELIGIONS	CURRENT AFTERLIFE BELIEFS OF MAJOR RELIGIONS			
AFTERLIFE BELIEF ELEMENT	**CHRISTIAN**	**MUSLIM**	**HINDU**	**BUDDHIST**
1. Belief in Afterlife/Eternal life	YES	YES	YES	YES
2. Belief in Universal Salvation	**NO**	**NO**	YES	YES
3. Belief in Final Judgment	YES	YES	YES	**NO**
4. Belief in Crown of Life (Ankh) and going to live with God	YES	YES	YES	YES
5. Belief in Savior (Christ/Horus)	YES	**NO**	**NO**	**NO**
6. Belief in **REINCARNATION**[a,b]	**NO**	**NO**	YES	YES
7. Belief in Hell/Hades/underworld	YES	YES	**NO**	**NO**
ADD-ON BELIEFS				
8. Belief in Eternal Damnation	YES	YES	**NO**	**NO**

[a] Modern Christians misinterpret reincarnation to be "born again," which means having the Spirit "reborn" in the same physical body during a single physical life lived in the one-lifetime hypothesis.
[b] Two small Muslim sects do believe in reincarnation.

BRINGING BACK THE MISSING PIECES TO CHRISTIANITY AND ISLAM: Today, bringing back into the Christian salvation myth and the Muslim salvation myth the knowledge that the near-death experience provides about transmigration of the consciousness (out-of-body reincarnation) and also the awareness that the life review includes no retribution would make Christianity and Islam nearly identical with the Ancient Egyptian religion, which is exactly what St. Augustine said in 427 AD:

> *"That which is called the 'Christian' religion existed among the ancients, and never did not exist, from the beginning of the human race until Christ came in the flesh, at which time the true religion WHICH ALREADY EXISTED began to be called Christianity."*

This acceptance of the restoration of reincarnation with no damnation would allow Christians more lifetimes to improve their spirituality, and the realization of future lifetimes in which Muslims could continue to improve their karma by following the five pillars of the faith would overcome any need to "HELP GOD" through jihad… Please, if God is actually almighty, then he does not need our mortal help. This also allows both Christians and Muslims to accept past-life memories.

APPENDIX B

.

SPIRITUALISM & QUANTUM ELECTRODYNAMICS: A COMING PARADIGM SHIFT

In the early 1850s, James Clerk-Maxwell developed the classical theory of electromagnetism. This breakthrough formed the basis of wireless (radio) communications technology, and underlies quantum theory. In fact, Albert Einstein believed that Maxwell's equations are basic to everything.

Later, in the latter half of the twentieth century, Dr. Richard Feynman felt that this one breakthrough, as quantified in Maxwell's equations, *"is the most important change in all of the history of science."*

Synergistically, this scientific advance of defining electromagnetic vibrations happened in parallel with the rise of Spiritualism (the belief in the possibility of communicating with departed souls continuing to exist at higher vibrations). Consequently, there are deep connections between these two separate philosophies.

Modern Spiritualism, as it is called by its proponents, began in 1848, at Hydesville, New York, when the Fox family was contacted by the disincarnate spirit of a salesman, Charles B. Rosna, who had, five years earlier, been murdered in the house they were now renting and was buried in the basement. In the basement, hair and teeth were found, but the rest of the body had apparently been moved. However, forty-plus years later, Rosna's bones and his salesman's pack were found behind a false basement wall in the foundation structure, which the murderer had constructed.

The Fox sisters soon began holding séances, receiving messages from many deceased persons. The truth of this concept spread rapidly, and within three decades, prominent scientists like **William James** of Harvard and **Sir Oliver Lodge** of England were carefully testing spiritualist mediums. Both of these men worked extensively with Lenora Piper in Boston and Gladys Osborne Leonard in London between the 1880s and 1920 (James died in 1910, and Lodge in 1947). Yet the rigorous scientific reports of these prominent scientists on the subject are fascinating to read even a hundred years later.

Both Lodge and James were convinced that the individual human consciousness (soul) does survive death and continues in some mode of existence capable of interacting with the living through mediumship.

Here is what Sir Oliver Lodge, a fellow of the Royal Society in 1887 and Knighted in 1902, said after studying these mediums regarding what science had proven about the afterlife:

> *"The first thing we learn is continuity* (life continues). *There is no such break in the conditions of existence as may have been anticipated* (no death except of the physical body)*; and no break at all in the continuous and conscious identity of genuine character and personality. Essential belongings, such as memory, culture, education, habits, character, and affection. All these, and to a certain extent tastes and interests...are retained* (in the afterlife).

> *"Meanwhile, it would appear that knowledge is not suddenly advanced, we are not suddenly flooded with new information, nor do we at all change our identity* (i.e. to become an angel or a saint)*; but powers and faculties are enlarged, and the scope of our outlook on the universe may be widened and deepened. If effort here has rendered the acquisition of such extra insight legitimate and possible."*

The above quotation is taken from Lodge's book *The Survival of Man*, showing that he formed his opinions of the reality of the afterlife from the scientific studies he had completed for the Society of Psychical Research.

Unfortunately, many people mistakenly believe that it was when Lodge's son, Raymond Lodge, died in the First World War that Lodge was overcome with grief and became a Spiritualist believer as a result, and that this emotional situation colored his science on the afterlife. This *excuse*, propagated by materialists, is false because Raymond died in 1915, six years after Lodge's 1909 book verifying consciousness survival was already published.

Raymond's coming through to Lodge in the mediumship of Gladys Osborne Leonard and Alfred Lord Peters was only a confirmation of the science that Oliver Lodge already knew, as he stated in his 1916 book, *Raymond or Life and Death,* in which he said:

> **"The number of more or less convincing proofs which we have obtained is by this time very great... I am as convinced of continued existence on the other side of death as I am of existence here... It may be said, you cannot be as sure as you are of sensory experience. I say I can. A physicist is never limited to direct sensory impressions; he has to deal with a multitude of conceptions and things for which he has no physical organ—the dynamical theory of heat, for**

instance, and of gases, the theories of electricity, of magnetism, of chemical affinity, of cohesion, aye, and his apprehension of the ether itself, lead him into regions where sight and hearing and touch are impotent as direct witnesses, where they are no longer efficient guides."

AN AKASHIC RECORD: One explanation popular a hundred years ago was that the souls of the departed are somehow permanently imprinted in an "ethereal medium" that surrounds and permeates all ordinary matter, which conveniently explains psychometry (where a person brings an object to the medium that belonged to the departed person and the medium "reads" the person's story off the object).

Today, physicists call this imprinting decoherence. The energy field where all this takes place is dark matter and dark energy. The astronaut Edgar Mitchell has recently described the "recording" as the *quantum hologram* that physically registers, by electromagnetic decoherence, everything that has ever happened. This recording process and reading the information is very similar to the information gathered by fMRI scans.

At about the same time that Modern Spiritualism was born (1848), **James Clerk Maxwell** began to conceive of electric and magnetic effects in a completely different way. Building on the earlier suggestions of Faraday, Maxwell—who is undoubtedly as great a physicist as Newton and Einstein because he combined electricity, magnetism, and light into one medium— also conceived of an all-embracing ether (dark matter) as being the mechanism and embodiment of the forces of electromagnetism.

Maxwell showed that this ethereal medium was capable of conveying energy in the form of electromagnetic waves propagating at the speed of light. Indeed, he surmised that light itself is an electromagnetic wave. Maxwell's final synthesis of these ideas was published in 1873, and in the 1880s, Hertz, who succeeded in producing and detecting them directly by means of oscillating electrical circuits, showed the reality of electromagnetic waves. In an article on "ether" for the Encyclopedia Britannica, Maxwell wrote:

> *"Ether or Aether: A material substance of a more subtle kind than visible bodies, supposed to exist in those parts of space which are apparently empty... Whatever difficulties we may have in forming a consistent idea of the constitution of the aether, there can be no doubt that the interplanetary and interstellar spaces are not empty, but are occupied by a material substance or body, which is certainly the largest, and probably the most uniform body of which we have any knowledge. Whether this vast homogeneous expanse of isotropic matter is fitted not only to be a medium of physical interaction between distant bodies, and to fulfill other physical functions of which, perhaps, we have as yet no conception,*

but also ... to constitute the material organism of beings exercising functions of life and mind as high or higher than ours are at present—is a question far transcending the limits of physical speculation."

Today, from the perspective of QED, it is obvious that Maxwell is attempting to describe dark energy and dark matter as explaining our *non-local* conscious universe.

But some of Maxwell's close friends and followers were less cautious in their ethereal speculations. For example, in 1873, Peter Guthrie Tait, along with Balfour Stewart, wrote a book entitled *The Unseen Universe,* expounding on the probable spiritual functions of the luminiferous ether.

"We attempt to show that we are absolutely driven by scientific principles to acknowledge the existence of an Unseen Universe, and by scientific analogy to conclude that it is full of life and intelligence—that it is in fact a spiritual universe and not a dead one."

The book (which enjoyed huge popularity in its time) argued that *"the forms of matter and mind survive eternally as configurations of spirit in the ether."*

Sir Oliver Lodge had written in another of his books, *The Ether of Space,* that:

"The universe we are living in is an extraordinary one; and our investigation of it has only just begun. We know that matter has a psychical significance, since it can constitute brain, which links together the physical and the psychical worlds. If anyone thinks that the ether, with all its massiveness and energy, has probably no psychical significance, I find myself unable to agree with him."

Quantum electrodynamics lends itself to various forms of neospiritualism based on the phenomenon of quantum entanglement (*non-locality*). Modern efforts to reconcile general relativity with quantum electrodynamics is partly motivated by the old yearning to apprehend the underlying structure and substance of existence and spirit.

Today, with the realization accepted by most quantum electrodynamic theoretical physicists that the universe is formed of conscious thought, rigorous science is finally filling in all the blanks that seemed previously to defeat the original Spiritualist theories of the late nineteenth century. Instead, the mounting evidence is actually proving that Spiritualism is indeed true, and the dream of the materialists that there is no soul (a religion in itself) is being completely undermined.

Now, it appears that in the twenty-first century, rapidly continuing over the next thirty years, there will be a reawakening of Spiritualism in the US and Canada (as well as Great Britain, Australia, and New Zealand) that will dwarf the tremendous growth this philosophy experienced in 1848–1888, and again in 1914–1927.

Rigorous triple-blind science is currently verifying the phenomenon of Spiritualism, and these discoveries are spreading across the Internet so rapidly that this is forcing a paradigm shift on the scientific community. By mid-century, any twenty-first-century scientist who still chooses to cleave desperately to the formerly comfortable Newtonian materialist theories of the nineteenth century will be seen as the old "quacks" that they truly are.

Bibliography

THE DEATH EXPERIENCE: What It Is Like When You Die. My 2012 book describes the story I have told today but without debunking Materialism or fully explaining QED, available in PAPERBACK or KINDLE from Amazon.

VIDEO 1 – "ON THE NATURE OF CONSCIOUSNESS & SURVIVAL" Interview with Dr. Alan Ross Hugenot (October 2015) 55 minutes. This interview covers most of this material in an entertaining way: available at https://www.youtube.com/watch?v=yByEQfaD314

VIDEO 2 – "SCIENCE OF THE AFTERLIFE" A September 2013 television interview with Dr. Hugenot speaking on the *Science of the Afterlife,* as delineated in his book *THE DEATH EXPERIENCE, What It Is Like When You Die,* is available at http://www.youtube.com/watch?v=sG8RAVh4VwE

VIDEO 3 – "BEYOND OUR SIGHT" A September 2014 film (57 minutes) in which Dr. Alan Hugenot discusses the near-death experience and mediumship, along with Dr. Dean Radin and four other mediums and NDE survivors is available at https://www.youtube.com/watch?v=xpSuO8DtiMM

VIDEO 4 – "Dr. Dean Radin" *Chief Scientist at Institute of Noetic Sciences (noetic.org) discusses consciousness studies at* Noetic.org *VIDEO CLIP (3 minutes) is available at* http://vimeo.com/113981492

VIDEO 5 – "Dr. Arnaud Delorme, *" mediumship scientist at Institute of Noetic Sciences* Noetic.org *VIDEO CLIP (3 minutes) is available at* http://vimeo.com/113345358

NUMEROUS OTHER VIDEOS ARE GROUPED AT THE CHAPTER ENDS OF EACH SUBJECT.

ERASING DEATH—The Science That Is Rewriting the Boundaries Between Life and Death, ©2013, by Dr. Sam Parnia, M.D. with Josh Young.

QUANTUM ENIGMA—Physics Encounters Consciousness, ©2011, by Bruce Rosenblum and Fred Kuttner, Oxford University Press.

CONSCIOUSNESS BEYOND LIFE—The Science of the Near-Death Experience, ©2010, by Pim van Lommel, M.D., HarperCollins.

BEST EVIDENCE 2d, ©2012, by Michael Schmicker.

DEMYSTIFYING THE OUT-OF-BODY EXPERIENCE, ©2012, by Luis Minero.

BIOCENTRISM—How Life & Consciousness Are Keys to Understanding the True Nature of the Universe, ©2009, by Dr. Robert Lanza M.D. with Bruce Berman.

THE BIOLOGY OF BELIEF—Unleashing the Power of Consciousness, Matter & Miracles, ©2005, by Bruce Lipton, Ph.D., Hay House, Inc.

HANDBOOK OF NEAR-DEATH STUDIES—30 Years of Investigation, ©2009, by Janice Miner Holden, EdD., Bruce Greyson, M.D., and Debbie James, RN/MSN, ABC-CLIO, LLC.

MESSAGES OF HOPE: The Metaphysical Memoir of a Most Unexpected Medium, ©2011, Suzanne Giesemann Former Commander USN, Aide de Camp to the Chairman of the Joint Chiefs of Staff.

INDUCED AFTER-DEATH COMMUNICATION, ©2005 & ©2014, by Dr. Allan Botkin.

The SURVIVAL OF MAN, ©1909, Sir Oliver Lodge.

RAYMOND OR LIFE AND DEATH, *©1916, Sir Oliver Lodge.*

End notes

1. Steven Weinberg, Nobel Laureate in Physics.

2. James Clerk Maxwell 9th ed. Encyclopedia Britannica.

3. Dr. Carl Sagan (1934–1996).

4. HUBRIS - Overbearing pride, presumption, and arrogance, which will eventually result in disastrous situations. It is defiance, arrogance, and thinking you can outsmart the gods.

5. *Series of Experiments Shed Light on the Role of Consciousness*, y Dr. Dean Radin, Ph.D. noetic.org/blog/dean-radin/double slip experiment.

6. Voltaire (Francoise-Marie Arouet 1694–1778).

7. *Das Wesen der Materie* [The Nature of Matter], speech at Florence, Italy (1944) (from Archiv zur Geschichte der Max-Planck-Gesellschaft, Abt. Va, Rep. 11 Planck, Nr. 1797).

8. Hank Wesselman, Ph.D.

9. *Science* 20 February 2015, vol 347 p. 6224, "That the speed of light in free space is a constant has been the cornerstone of modern physics. However, light beams have a finite transverse size, which leads to a modification of their wave vectors resulting in a change to their phase and group velocities. Our work highlights that, even in free space, the invariance of the speed of light only applies to plane waves."

10. *NATURE* in 2010 *"Cycling at the Speed of Light."*

11. David Chalmers, p. 4, *Consciousness and Its Place in Nature.*

12. Robert Almeder, ©1992 Ph.D. professor of philosophy, Georgia State University.

13. Willis W. Harman of the Institute of Noetic Sciences (noetic.org).

14. Max Planck, *Scientific Autobiography and Other Papers.*

15. Dr. Gary Schwartz, Ph.D., *Consciousness, Spirituality and Postmaterialist Science: An Empirical and Experiential Approach,*

© 2012 Lisa Miller (2012) Oxford Handbook of Psychology and Spirituality.

16. Lipton, Bruce H. Ph.D. *The Biology of Belief,* ©2005.

17. Lipton, Bruce H. Ph.D. & Steve Bhaerman, *Spontaneous Evolution: Our Positive Future* ©2009.

18. p. 148, *The French Revelation,* Edward C. Randall's Complete Works Compiled by N. Riley Heagerty ©1995.

19. Dr. David Hossack, professor at Columbia University (King's College) and founder of the Medical School, graduated from Princeton and received his medical degree in Philadelphia in 1791, then studied abroad at Edinburgh, Scotland, before returning to New York. He was one of the founders of the New York Historical Society in 1804. He also established the Elgin Botanical Garden between 47th and 51st streets and between 5th and 6th.

20. p. 149, *The French Revelation*, Edward C. Randall's Complete Works Compiled by N. Riley Heagerty ©1995.

21. p. 184, *Light & Death* ©1998 Dr. Michael Sabom.

22. *Gospel of Phillip*, (Nag Hammadi Library Codex II Tractate 3 p.61).

23. *A Course in Miracles*, Chapter 5, Section VI ©1977.

24. p. 39 *Coming Back to Life: The After-Effects of the Near-Death Experience* ©2001 P.M.H. Atwater, Balantine Books.

25. St. Paul's letter to the *Ephesians 2:6-9* (NIV).

26. *Life After Life*. by Raymond L. Moody Jr. M.D. ©1975, 1988 Bantam edition.

27. Ibid.

28. St. Paul's letter *1 Corinthians 2:4-5* (NIV).

29. p. 75, *Heading Toward Omega* ©1984, Kenneth Ring.

30. *Life After Life* by Raymond A. Moody ©1975. And a translation by Paul Shorey, in Hamilton and Cairns (ed.), Plato: *The Collected Dialogues* (New York Bollingen Series LXXI, 1961), pp. 838-40.

31. *The Jesus Mysteries* ©1999 Timothy Freke & Peter Gandy.

32. St. Augustine, *Retractions*, taken from p. 54 *The Fathers of the Church, Saint Augustine, The Retractions*. Brogan c.1968 Catholic University of America Press. Augustine's *Retractions* were written as a clarification in 428 c.e., after publication of his seminal work *The City of God* in 427c.e. *My own copy of St. Augustine's text, published by the Catholic University of America Press, carries the Nihil Obstat of John C. Selner, S.S., S.T.D. the Censor Librorum, and also the Imprimatur of Patrick Cardinal A. O'Boyle, D.D. Archbishop of Washington. These are both official declarations that a book or pamphlet is **FREE OF DOCTRINAL OR MORAL ERROR**. No implication is contained therein that those who have granted the Nihil Obstat and Imprimatur agree with the contents, opinions, or statements expressed. In laymen's terms, this means that the Catholic Church officially agrees that these are the actual words of Saint Augustine himself, and they agree that he actually said, "That which is called the Christian religion existed among the ancients, and never did not exist, from the beginning of the human race." Saint Augustine clearly admits that paganism was in fact the Christian religion. Consequently, seeing the Roman emperor Justinian condemn paganism (Origenism) 120 years later illustrates that the emperors had no desire for true religion, only power and control. That original ancient religion St. Augustine identified as the true faith allowed for second chances without hellfire.*

33. *Twenty Cases Suggestive of Reincarnation*, by Ian Stevenson, M.D. ©1974 University of Virginia.

34. *Children Who Remember Previous Lives*, by Ian Stevenson, M.D., ©2001 McFarland & Company, Inc.

35. *Life Before Life*, by Jim B. Tucker, M.D. ©1995.

36. p. 183, *Twenty Cases Suggestive of Reincarnation*, by Ian Stevenson, M.D. ©1974 University of Virginia.

37. p. 5, *Death and Personal Survival*, by Robert Almeder, ©1992 Rowman & Littlefield Publishers, Inc.

38. *Reincarnation: Verified Cases of Rebirth After Death*, by K.K.N. Sahay ©1927 N.L. Guipta.

39. p. 183, *Immortal Remains: The Evidence for Life After Death*, by Stephen E Braude © 2003.

40. p. 19, *Death and Personal Survival*, by Robert Almeder, ©1992 Rowman & Littlefield Publishers, Inc. Almeder also lists numerous other publications that discuss this case uncritically.

41. p. 68, *Children Who Remember Previous Lives*, by Ian Stevenson, M.D., ©2001 McFarland & Company, Inc.

42. *Soul Survivor, The Reincarnation of a World War II Fighter Pilot* ©2010 by Bruce & Andrea Leininger.

43. *Jewish Tales of Reincarnation* by Yonassan Gershom ©1999.

44. *Beyond The Ashes: Cases of Reincarnation from the Holocaust,* by Yonassan Gershom ©1992.

45. *From Ashes to Healing: Mystical Encounters with the Holocaust,* by Yonassan Gershom ©1996.

46. p. 170, *Human Personality and Its Survival of Bodily Death,* by F.W.H. Myers ©1903, 1961, 2001.

47. *Life After Death* by Bill Newcott p. 66 Sept / Oct 2007 *AARP* Magazine.

48. p. 87, *The Heart of the Mind* by Jane Katra, Ph.D. and Russell Targ ©1999, Reuter's News Service January 31, 1998.

49. p. 517-524, *Proceedings: Society for Psychical Research #36,* ©1927, *Parapsychology, A Century of Inquiry,* D. Scott Rogo ©1975.

50. p. 187, *Ghost Hunters* by Deborah Blum, ©2006, *Mediumship & Survival* by Alan Gauld ©1983.

51. p. 275-285, *Ghost Hunters* by Deborah Blum, ©2006.

52. *Natural and Supernatural: A History of the Paranormal,* Brian Inglis ©1977, 1992.

53. p. 92, *The Heart of the Mind* by Jane Katra, Ph.D. and Russell Targ ©1999; "The Case of Runolfur Runolfsson" p. 33-59 *Journal of American Society for Psychical Research* No. 69 (1977).

54. *Survival Is in the Details: Emerging Evidence for Discarnate Intention from Mediumship Research,* Gary E. Schwartz, Ph.D. and Julie Beischel, Ph.D. ©2006 University of Arizona.

55. Ibid.

56. p. 201, *Beyond the Ashes,* by Yonassan Gershom ©1992.

57. p. 173, *The French Revelation,* Edward C. Randall's Complete Works Compiled by N. Riley Heagerty ©1995.

58. Dr. David Hossack, professor at Columbia University (King's College) and founder of the Medical School, graduated from Princeton and received his medical degree in Philadelphia in 1791, then studied abroad at Edinburgh, Scotland, before returning to New York. He was one of the founders of the New York Historical Society in 1804. He also established the Elgin Botanical Garden between 47th and 51st streets and between 5th and 6th.

59. Randall p. 150-151.

60. p. 15, *The French Revelation,* Edward C. Randall's Complete Works Compiled by N. Riley Heagerty ©1995.

61. p. 148, *The French Revelation,* Edward C. Randall's Complete Works Compiled by N. Riley Heagerty ©1995.

62. Dr. David Hossack, professor at Columbia University (King's College) and founder of the Medical School, graduated from Princeton and received his medical degree in Philadelphia in 1791, then studied abroad at Edinburgh, Scotland, before returning to New York. He was one of the founders of the New York Historical Society in 1804. He also established the Elgin Botanical Garden between 47th and 51st streets and between 5th and 6th.

63. p. 149, *The French Revelation,* Edward C. Randall's Complete Works Compiled by N. Riley Heagerty ©1995.

64. *Across Time and Death,* ©1993 Jenny Cockell, Fireside Books, London.

65. *Other Lives, Other Selves,* by Roger J. Woolger, Ph.D. © 1988.

66. *Miracles of Mind* by Jane Katra, Ph.D. and Russell Targ ©1998; *The Heart of the Mind* by Jane Katra, Ph.D. and Russell Targ ©1999; *Limitless Mind* by Russell Targ ©2004.

67. *Extraordinary Knowing* by Elizabeth Lloyd Mayer, Ph.D. ©2007 Published posthumously.

68. *Psi Conducive States*, William G. Braud 1975, *Journal of Communications* 25 p-142-152.

69. *Extraordinary Knowing* by Elizabeth Lloyd Mayer, Ph.D. ©2007 Published posthumously.

70. STAPP, H. *Compatibility of Contemporary Physical Theory with Personality Survival*, 2010.

71. LASLOW, E. *New Concepts of Matter, Life and Mind.*

72. *Extra Sensory Perception of Quarks*, Phillips, S.M. ©1980 Quest Books, Wheaton IL.

73. Feynman, R. Lectures on Physics, Chapter 37, Quantum Behavior 31-1 Atomic Mechanics.

74. Heagerty, N. Riley, *The French Revelation* (c)1995 p. 15.

75. Heagerty, N. Riley, *The French Revelation* (c)1995 p. 229.

76. Heagerty, N. Riley, *The French Revelation* (c)1995 p. 230.

77. PLANCK, M. *Das Wesen der Materie* [The Nature of Matter], speech at Florence, Italy (1944) (from Archiv zur Geschichte der Max-Planck-Gesellschaft, Abt. Va, Rep. 11 Planck, Nr. 1797).

78. LANZA, R. "Biocentrism: How Life and Consciousness Are the Keys to Understanding the Universe."

79. VON NEUMANN, J. (1955/1932) Mathematical Foundations of Quantum Mechanics. Princeton University Press.

80. TOMONAGA, S. (1946) "On a relativistically invariant formulation of the quantum theory of wave fields." *Progress of Theoretical Physics*, 1,27-42.

81. SCHWINGER, J. (1951) "Theory of Quantized Fields 1." *Physical Review*, 82, 914-927.

82. LANZA, R. Ibid.

83. Quoted from p. 21, *My View of the World* by Erwin Schrödinger, ©1964 Cambridge University Press, London.

84. Quoted from p. 43-44, *The Phenomenon of Man* by Teilhard de Chardin, ©1965 Harper Torchbooks.

85. The coffin texts from the Nicropolis at Saqquara (2450 BC).

86. *The Jesus Mysteries* ©1999 Timothy Freke & Peter Gandy.

87. St. Augustine, *Retractions*, taken from p. 54 *The Fathers of the Church, Saint Augustine, The Retractions*. Brogan c.1968 Catholic University of America Press. Augustine's *Retractions* were written as a clarification in 428 c.e., after publication of his seminal work *The City of God* in 427 c.e. *My own copy of St. Augustine's text, published by the Catholic University of America Press, carries the Nihil Obstat of John C. Selner, S.S., S.T.D. the Censor Librorum, and also the Imprimatur of Patrick Cardinal A. O'Boyle, D.D. Archbishop of Washington. These are both official declarations that a book or pamphlet is **FREE OF DOCTRINAL OR MORAL ERROR**. No implication is contained therein that those who have granted the Nihil Obstat and Imprimatur agree with the contents, opinions, or statements expressed. In laymen's terms, this means that the Catholic Church officially agrees that these are the actual words of St. Augustine himself, and they agree that he actually said, "That which is called the Christian Religion existed among the ancients, and never did not exist, from the beginning of the human race." Saint Augustine clearly admits that paganism was in fact the Christian religion. Consequently, seeing the Roman emperor Justinian condemn paganism (Origenism) 120 years later illustrates that the emperors had no desire for true religion, only power and control. That original ancient religion St. Augustine identified as the true faith allowed for second chances without hellfire.*

88. p. 226-227 "Justinian intervened in ecclesiastical matters more forcefully and systematically than any of his predecessors"... "With vigor he sought to wipe out the not inconsiderable remnants of paganism in the East"..."Backsliders into paganism were to be put to death"...."He even interfered to dictate Jewish doctrine forbidding teaching against the Last Judgment, resurrection of the dead and existence of angels. Rabbis were compelled to allow

reading of the Bible in the synagogues in Greek or Latin along with Hebrew... P. 246 Scholars are not agreed on how these anathemas became connected with the Council... Some say the bishops gathered for the Council approved the anathemas presented to them by the emperor before the actual opening of the official proceedings... At any rate they are continually linked to the work of the Council." *The First Seven Ecumenical Councils (325–787) Their History and Theology* ©1983, Leo Donald Davis. *This is a great read, and understanding the underlying regional differences in Christianity between Byzantium—Rome and Africa (Carthage-Alexandria) is very important in comprehending how the emperors played the factions against each other, and also that the doctrinal Christianity that has been handed down to us by this history is indeed the watered-down version the Roman emperors needed to control the populace. Justinian did not murder "heretics" because he was a zealous Christian; it was instead because he was consolidating his power over his subjects.*

89. p. 79 *Old Souls* by Tom Schroder ©1999 Tom Schroder.

CPSIA information can be obtained
at www.ICGtesting.com
Printed in the USA
LVOW06s1722120617
537839LV00016B/39/P

9 781457 546945